W9-CJY-388

CONSCIOUSNESS REGAINED

CONSCIOUSNESS REGAINED

Chapters in the development of mind

NICHOLAS HUMPHREY

Oxford New York
OXFORD UNIVERSITY PRESS
1983

Oxford University Press, Walton Street, Oxford OX2 6DP
London Glasgow New York Toronto
Delhi Bombay Calcutta Madras Karachi
Kuala Lumpur Singapore Hong Kong Tokyo
Nairobi Dar es Salaam Cape Town
Melbourne Auckland
and associates in
Beirut Berlin Ibadan Mexico City Nicosia

Oxford is a trade mark of Oxford University Press

British Library Cataloguing in Publication Data
Humphrey, Nicholas
Consciousness regained.
1. Consciousness 2. Genetic psychology
I. Title
153 BR311
ISBN 0-19-217732-X

Library of Congress Cataloging in Publication Data
Humphrey, Nicholas.
Consciousness regained.
Bibliography: p.
Includes index.
1. Consciousness – Addresses, essays, lectures.
2. Genetic psychology – Addresses, essays, lectures.
3. Psychology – Addresses, essays, lectures.
I. Title.
BF311.H775 1983 155.7 82-24527
ISBN 0-19-217732-X

Photoset by Rowland Phototypesetting Ltd
Printed in Great Britain by
Thomson Litho Ltd
East Kilbride, Scotland

ACKNOWLEDGEMENTS

Chapters 1 and 6–8 appear here for the first time. The other chapters are based on published articles and broadcasts, but have been partly rewritten (and in some cases retitled). The list below gives details of published articles which are close to the present text. I am grateful to the original publishers for their permission to include versions of these articles.

'The social function of intellect': in *Growing Points in Ethology*, ed. P. P. G. Bateson and R. A. Hinde, Cambridge University Press, 1976.

'Nature's psychologists': based on the Lister Lecture for the British Association for the Advancement of Science, Birmingham, 1977; broadcast on Radio 3 in 1979; in *Consciousness and the Physical World*, ed. B. Josephson and V. Ramachandran, Pergamon Press, 1980.

'Having feelings, and showing feelings': based on a paper for a workshop on 'Self-awareness in domesticated animals', Oxford, 1980; in *Self-awareness in Domesticated Animals*, ed. D. G. M. Woodgush, M. Dawkins and R. Ewbank, Universities Federation for Animal Welfare, 1981.

'Consciousness: a Just-So Story': based on a lecture for the International Congress of Ethology, Oxford, 1981; *New Scientist*, 95, 474, 1982.

'The illusion of beauty': based on a lecture at the Institute of Contemporary Arts, London, 1973; broadcast on Radio 3 in 1979; *Perception*, 2, 429, 1973.

'Turning the left cheek': originally 'Status and the left cheek', with Christopher McManus, *New Scientist*, 59, 437, 1973.

'Butterflies that stamp': originally 'Variations on a theme', *New Scientist*, 63, 233, 1974.

'The colour currency of nature': in *Colour for Architecture*, ed. T. Porter and B. Mikellides, Studio-Vista, 1976.

'Contrast illusions in perspective': *Nature*, 232, 91, 1971.

'An ecology of ecstasy': review of *The Spiritual Nature of Man* by Alister Hardy (Oxford, 1979), *London Review of Books*, 17 April 1980.

'Straw ghosts': review of *This House is Haunted* by G. L. Playfair (London, 1980), and *Science and the Supernatural* by John Taylor (London, 1980), *London Review of Books*, 2 October 1980.

'Karma is raining on my head': originally 'New ideas, old ideas', review of *Mind and Nature: A Necessary Unity* by Gregory Bateson (London, 1979), *London Review of Books*, 6 December 1979.

'What is mind? No matter. What is matter? Never mind': review of *The*

Mind's I, ed. Douglas Hofstadter and Daniel Dennett (Hassocks, 1981), *London Review of Books*, 17 December 1981.
'An immodest proposal': prepared as a radio broadcast; *Sanity*, June 1982.
'Four minutes to midnight': the 1981 Bronowski Memorial Lecture, broadcast on BBC 2 television, October 1981; BBC Publications, 1981.

Acknowledgement is due for the use of illustrations as follows:
Figure 1, Deutsches Museum, Munich; Figure 3, *American Journal of Psychology*, University of Illinois Press; Plate 2, The Fine Arts Museums of San Francisco; Plate 3, Städelsches Kunstinstitut, Frankfurt (photo: Ursula Edelmann); Plate 4, National Museum Vincent van Gogh, Amsterdam; Plate 5, Stichting Johan Maurits van Nassau Mauritshuis; Plate 6, Peking Palace Museum; Plate 7, The Trustees, The National Gallery, London.

PREFACE

> The first and wisest of them all professed
> To know this only, that he nothing knew.
>
> John Milton, *Paradise Regained*

For some time I have been meaning to write a real book. It would have brought together my ideas about the evolution of consciousness, about art, culture and politics, in the form of a single extended essay with a beginning, a middle and an end. It would have been a carefully balanced work of scholarship, with all criticisms anticipated and all counter-arguments deflected. This is not that book.

Over the years I have in fact written more beginnings, middles and ends than I care to remember. But the pieces, instead of forming parts of a grand whole, seemed always to be pieces still. Some I gave as one-off lectures, others I published as short articles or reviews, others I simply put in a drawer. If Henry Hardy, at Oxford University Press, had not come to the rescue with his suggestion of a book made up of separate essays, the project would probably have died. The 'real' book, he realised, had run into the ground; but another book, just as representative of my ideas, more polemical maybe but for that very reason potentially more readable, was already in existence.

The theme that runs through this collection is a concern with *why* human beings are as they are. Why has consciousness evolved, why are human beings so good (or sometimes so bad) at reading other people's minds, why do human beings dream, why love beauty, why believe in ghosts . . . why acquiesce in preparations for mass suicide? The answers that most interest me are historical and evolutionary. Human beings are as they are because their history – I mean their history for at least the last five million years – has been (so we may guess) as it has been.

In so far as I have drawn for this book on material that has previously been published or broadcast, I have kept more or less to the original texts. That means there is a generous overlap between certain chapters, and no overlap at all between others. There is also less unity of style than there might otherwise have been. In compensation, it means that every chapter can be read in isolation and, except perhaps for the new material in Chapters 6–8, in any order.

I have written and thought about these subjects during interesting times. My academic base has been the Sub-Department of Animal Behaviour at Cambridge University, where its directors, Patrick Bateson and Robert Hinde, have all along provided intellectual and moral support. At home I have had the help and encouragement of Caroline Humphrey, Susannah York, and my parents John and Janet Humphrey. One of the essays, on 'Turning the Left Cheek', was written with Christopher McManus, who (while still a student at Cambridge) collected all the data for it. Henry Hardy has been the best of editors.

CONTENTS

FOUR MINUTES TO MIDNIGHT

ILLUSTRATIONS

TEXT FIGURES

PLATES

(between pages 146 and 147)

NATURAL PSYCHOLOGY AND THE EVOLUTION OF CONSCIOUSNESS

How do I know the way of all things?
By what is within me.

Lao-tzu, *Tao-te-ching*, 6th century BC

By what evidence do I know, or by what considerations am I led to believe, that there exist other sentient creatures; that the walking and speaking figures which I see and hear, have sensations and thoughts, or in other words, possess Minds? . . . first, they have bodies like me, which I know in my own case, to be the antecedent condition of feelings . . . secondly, they exhibit the acts and other outward signs, which in my own case I know by experience to be caused by feelings. I am conscious in myself of a series of facts connected by an uniform sequence, of which the beginning is modifications of my body, the middle is feelings, the end is outward demeanour. In the case of other human beings I have the evidence of my senses for the first and last links of the series, but not for the intermediate link. I find, however, that the sequence between the first and last is as regular and constant in those other cases as it is in mine. In my own case I know that the first link produces the last through the intermediate link, and could not produce it without. Experience, therefore, obliges me to conclude that there must be an intermediate link; which must either be the same in others as in myself, or a different one: I must either believe them to be alive or to be automatons: and by believing them to be alive, that is, by supposing the link to be of the same nature as in the case of which I have experience, and which is in all other respects similar, I bring other human beings, as phenomena, under the same generalisations which I know by experience to be the true theory of my own existence.

John Stuart Mill, *An Examination of Sir William Hamilton's Philosophy*, 1865

1
INTRODUCTION: *HOMO PSYCHOLOGICUS*

Within little more than a week of the Creation, Eve had been beguiled by a subtle serpent, she had tempted Adam, and God Himself had been caught telling lies. 'But of the tree of the knowledge of good and evil,' God had said, 'thou shalt not eat of it: for in the day that thou eatest thereof thou shalt surely die.' But the serpent had told the woman, 'Ye shall not surely die.' And Eve had eaten the apple – and she had not died, nor had Adam. Men, women and Gods too, it seems, were deceivers ever.

The descent of man and his companions, having started with the Fall, evidently gathered pace with each new generation. The first books of the Bible are a chronicle of deceit, treachery, selfishness – of the cunning exploitation of one person by another. 'Subtlety', the characteristic of the serpent, finds its fullest expression in the work of human beings. Jacob, the smooth man, succeeds in passing himself off as his hairy brother Esau: 'Thy brother', says Isaac, 'came with subtlety and hath taken away thy blessing'. Jonadab tells Amnon, the son of David, how to seduce his sister by feigning sickness and then forcing her to bed when she comes to minister to him: 'Jonadab', we are told, 'was a very subtle man.' Solomon warns the youth of his country to beware of loose women, who despite their good looks are 'subtle of heart'.

The Bible may not be a reliable guide to human evolution. But its authors diagnosed one of the fundamental traits which distinguish Man in nature. Human beings are born psychologists. Subtle of heart and of head, they are uniquely skilful in their ability to handle one another. They know better than any other animal how to anticipate – and work upon – the behaviour of fellow members of their species.

But although, in reality as much as in fable, people will use this skill to promote their own interests over those of others, man's subtlety is not inherently a hostile trait. Those who have the psychological skill to be subtly exploitative have the skill to be subtly loving, subtly charitable and subtly altruistic too. And if, in the course of evolution, it has sometimes been to people's advantage to get the better of their fellow human beings, it has equally been to their advantage to get along with them: people have required their art as psychologists in order to humour, to reassure and to succour their allies, quite as often as they have required it to outwit their rivals.

It was, as I argue in succeeding chapters, the circumstances of primitive man's *social* life – his membership of a complexly interacting human community, his need to do well for himself while at the same time sustaining others – which did more than anything to make man, as a species, the subtle and insightful creature we know today. For, with the continual pressure towards greater mutual understanding, natural selection favoured two parallel developments in the evolution of the human mind.

'Social intelligence' required, for a start, the development of certain abstract intellectual skills. If men were to negotiate the maze of social interaction it was essential that they should become capable of a special sort of forward planning. They had to become *calculating* beings, capable of looking ahead to yet unrealised possibilities, of plotting, counter-plotting and pitting their wits against group companions no less subtle than themselves. Never before, in their dealings with the non-social world, the world of sticks and stones, not even in their dealings with the world of living predators and prey, had human beings needed the powers of abstract reasoning which they now needed in their dealings with each other. But now their very survival within the social group depended on it. And it asked, I believe, for a level of intelligence unmatched in any other sphere of living. *Homo*, once he threw in his lot with society, had given himself no choice but to become *Homo habilis*, next *Homo sapiens*. Man the clever, man the wise.

But cleverness alone was not enough. Before human beings could even begin to calculate where their own and others' behaviour would take them, it was essential that they should acquire a much deeper understanding of the character of the strange creature who stood at the centre of their calculations – Man himself. They had to have a way of finding out what men as such are like, how they react, what makes them tick. They had to become sensitive to other people's moods and passions, appreciative of their waywardness or stubbornness, capable of reading the signs in their faces and equally the lack of signs, capable of guessing what each person's past experience holds hidden in the present for the future. They had above all to make sense of the enigma of the ghost in the machine. In short they had to become 'natural psychologists'. Clever man had to become *Homo psychologicus*.

The emergence of man to the full status of natural psychologist has come about as a result of both biological and cultural evolution. It has required fundamental changes to the human brain, and later, as I discuss in chapters 6–8, the creation of special institutions within human culture. It has taken a long time. But it has won for the human species a remarkable – and puzzling – prize. Puzzling, because the ability to do psychology, however much it may nowadays be an ability possessed by every ordinary man and woman, is by no means an ordinary ability.

Let no one pretend that natural psychology – or psychology under any other title – is anything but an extraordinarily difficult thing to do. Philosophers and scientists who, with all the paraphernalia of theory and experimental method at their disposal, have been trying for a century or more to develop their own science of human behaviour, have found the task a daunting and a humbling one. Indeed academic psychology, as studied in the universities, has proved to be the most intractable branch of all the sciences. Psychology, both in theory and in practice, is much more difficult than physics.

I have at hand a textbook of Physical Mechanics. 'Consider', it says, 'the case of a system of bodies, attracting or repelling each other or acting on one another by contact or through connections . . .'.[1] It is of course just such considerations which must have been exercising the minds of every ordinary man and woman for several million years. Yet for the student of physics the 'bodies' in question are lifeless lumps of matter, attracted by the force of gravity, acting on one another by friction, connected by rods or strings; while for ordinary people they are the bodies of other human beings, attracted by sentiment, interacting through speech and gesture, connected by bonds of blood or friendship . . . and these, by the standards of physics, must be counted amongst the most lawless and unprincipled bodies in the universe.

There are not, and never will be, Newtonian principles of human behaviour. Those academic psychologists who have tried to emulate the method and theory of classical physics – who have tried like Clark Hull in the 1930s to write a latter-day *Principia* – have proved what any layman might have told them at the start: the mountain of human complexity cannot be turned into a molehill of scientific laws.

How is it, then, that human beings have acquired a *natural* ability to do psychology? How can the blind forces of evolution have succeeded, where objective scholarship has failed?

To answer a riddle with a paradox: blindness to theoretical objections may have provided the best chance of practical success. From the beginning, the task for the natural psychologist was no more and no less than this: that he should come up with a system for interpreting and predicting human behaviour which gave the right answers, for whatever reasons. Ignorant of the method and manners of the scientist, unaware of the warnings of any philosophical Cassandra, he had no choice but to adopt a policy of shameless pragmatism. If the job was worth doing, it was no doubt worth doing well. But given that it was the ends which mattered, not the means, he was never under any constraint to make his system of psychology theoretically respectable. If the job could not be done straightforwardly, it was worth doing deviously. And so, along the line, human beings were at liberty to adopt any style, any trick of reasoning which brought them nearer to their goal. If it sometimes meant their trafficking with techniques and ideas which flew in the face of what we now call objective scientific logic, so be it. What was wanted was a plain man's guide.

Not that the plain-man's guide, as it emerged, was all that plain. Indeed the formula for understanding human behaviour which has come to lie at the heart of natural psychology is arguably as fancy as it is philosophically impertinent. My thesis – I shall call it such for the time being – is that Nature's solution to the problem of doing psychology has been to give to every member of the human species both the power and inclination *to use a privileged picture of his own self as a model for what it is like to be another person*.

In short, what the natural psychologist is empowered to do is to enter by the light of his subjective experience into other people's minds. And in doing so he trusts to a principle which was stated with characteristic bluntness by Thomas Hobbes:

> [Given] the similitude of the thoughts and passions of one man, to the thoughts and passions of another, whosoever looketh into himself, and considereth what he doth, when he does *think*, *opine*, *reason*, *hope*, *fear*

&c. and upon what grounds; he shall thereby read and know, what are the thoughts and passions of all other men upon the like occasions.[2]

I shall come, in Chapter 6, to evidence that this is in fact what human beings unhesitatingly do, and that it works. Even allowing that there can never truly be a perfect match between one man and another – allowing that each individual has, as George Eliot commented in *Middlemarch*, 'an equivalent centre of self, whence the light and shadows must always fall with a certain difference' – none the less Hobbes's principle serves the psychologist better than any other principle known either to common sense or to science.

But first we should consider just what this way of doing psychology involves. On the face of it, the idea of using oneself as a model for others may perhaps seem nothing special. Any student of human behaviour has to start somewhere. Given that the natural psychologist is fortunate enough to be in his own right one of the very creatures whose behaviour he wants to understand, what could be more reasonable than that he should start by making observations on himself? His own body, besides being the human body with which of necessity he spends the greatest time, is a body with which he has a uniquely intimate relationship: he can observe it in secret and in the open, in sickness and in health, in the company of friends, enemies, parents, children, lovers . . . Scarcely surprising therefore that he should begin his analysis of other people by drawing on *self*-observation for most of his evidence about how a typical human being behaves. Even physicists have been known to refer to the evidence of their own bodies: it may not have been his own body which Galileo dropped from the Leaning Tower of Pisa, but it was his own body which Archimedes used to displace the water in the bath.

If that was all there was to it – if self-observation meant simply observation in the usual sense of 'observation', namely looking at a body from outside – natural psychology would indeed be nothing special. But to the natural psychologist, self-observation means not merely observing from outside but observing from within, not merely looking *at* one's own behaviour but looking *in* on it – in on the 'thoughts and passions' which accompany it. And the capacity for that kind of inner observation – the capacity to look into oneself and consider what one doth when one does think, hope, fear &c. – is something of an altogether different order. It represents, I believe, the most peculiar and sophisticated development in the evolution of the human mind.

To give it a name, it is the capacity for 'reflexive consciousness': consciousness of consciousness. As a biological capacity, I believe that it has evolved expressly to meet the exceptional needs of man as a psychologist; and I suspect it has no parallel in lower animals. Though it might be argued (I suggest as much myself in Chapter 3) that there are certain other highly social mammals – wolves, dolphins, maybe the anthropoid apes – whose way of life is sufficiently complex for them too to have developed some psychological skills, I know of no reason to suppose that these or any other species have in fact travelled the same road as human beings.

*

When the natural psychologist looks into himself, what does he find there? A 'beetle' in a box, was Ludwig Wittgenstein's reply: a beetle which only the subject himself can look at, which he can never compare with anyone else's beetle, and which – since it defies public definition – there is not much point in trying to discuss out loud. 'One can "divide through" by the thing in the box; it cancels out, whatever it is.'[3] Of this thing in the box, then, of inner experience in general, perhaps I ought to say no more . . .

But the fact is that, whatever may be the logical problems of describing inner experience, human beings everywhere openly attempt it. There is, so far as I know, no language in the world which does not have what is deemed to be an appropriate vocabulary for talking about the objects of reflexive consciousness, and there are no people in the world who do not quickly learn to make free use of this vocabulary. Indeed, far from being something which baffles human understanding, the open discussion of one's inner experience is literally child's play to a human being, something which children begin to learn before they are more than two or three years old.[4] And the fact that this common-sense vocabulary is acquired so easily suggests that this form of description is natural to human beings precisely because it maps directly on to an inner reality which each individual, of himself, innately knows. It is not necessarily the case, as Wittgenstein would have it, that all languages are 'games' for which the rules must be publicly agreed (with the corollary that where there can be no public verification there can be no language and no sense): the rules of the game may, as in this case, be written on the lid of each man's box.

It is true of course that different individuals will find different ways of expressing what their experience is like, and that some are more articulate than others. It is also true that some people may hold an unusual view of where, in relation to their bodies, the experience is located: the Dinka, an African tribe living in the southern Sudan, are said to regard some attributes of the 'inner self' as fields of consciousness outside their bodies.[5] None the less, when allowance is made for certain eccentricities, there is a remarkable convergence in the accounts which people of all races and all cultures give of what reflexive consciousness reveals to them. The gist of it – and I am attempting here to summarise, not to caricature – is this:

'In association with my body there exists a spirit, conscious of its own existence and its continuity in time. This is the spirit (mind, soul . . .) which I call "I". Among the chief attributes which "I" possess are these: I can act, I can perceive, and I can feel.

'Thus it is "I" who, by the exertion of my will, bring about almost all my significant bodily actions: I will my arm to rise – it rises; I will my lips to speak – they speak . . . It is "I" who, by means of my external senses, perceive the outside world: I see sights, hear sounds, smell smells – and so build up a picture of what is happening around me . . . And it is "I" who, within the boundaries of myself, feel states of emotion, sensations, moods and passions: I who am in pain, I who am scared, I who am consumed by jealousy . . .

'But over all this, "I" as a spirit have wants and aspirations. And those things I want or aspire to are largely dictated by my present or anticipated emotions, sensations, moods and passions. When I am in pain I want to ease it, when I am scared I want to find security, when I am jealous I want to take revenge . . .

'Functioning as "I" do as a unitary being, I work like this: by planning my actions in relation to what my perceptions tell me about their probable effects, I try to satisfy whatever wants or aspirations my states of feeling have aroused.'

Now all this is very odd. Indeed if this kind of story about one's inner self were not already so familiar to us – if it were not already in our bones – it might well seem no more than an elaborate fantasy. The idea of a conscious spirit which wills, feels, wants etc.

is seemingly the stuff of myth and metaphysics, not of science. Such notions have no place in Newton's, or any other materialist's, picture of the universe, and their objective status is to say the least uncertain. No wonder, perhaps, that there have been academic scientists and philosophers ready to declare that we should not be taken in: that the whole story is meaningless, that consciousness has no biological function and is merely an epiphenomenon, the irrelevant 'noise of the machine'. No wonder that there have been other scholars, led by the self-styled behaviourists, who have taken an even dimmer view, stoutly maintaining not only that the story is a fantasy, but that since it is essentially irrational, illogical, unverifiable, unfalsifiable and metaphysical, it must also be dangerously misleading.

Yet whatever else it is, common sense and common observation tell us that the story cannot in practice be misleading. If it were, if it led too often to false conclusions about the way that human beings behave, the story-tellers would surely have received their quietus at the hands of natural selection long ago. And yet today, wherever we look, we find members of the human species making free use of this story for the interpretation of behaviour. The proof of the apple has been in the eating: it has been eaten countless times in the course of human evolution, and men as psychologists have evidently thrived. It is no good the behaviourists' saying it can't be done. Wittgenstein was an aeronautical engineer before he turned philosopher, no doubt a clever one. But even he would surely have been laughed to scorn if, after examining a bird's wing, he had emerged from his study to announce that he had proved from first principles that it is logically impossible for birds to fly.

*

'The truth', Robert Pirsig wrote, 'comes knocking on the door and you say "Go away, I'm looking for the truth", and so it goes away.'[6] We should, I think, open the door to a straightforward explanation of why the story of the inner self evidently works so well. It is that, in the course of human evolution, natural selection has ensured that the description which reflexive consciousness gives of inner experience is anything but a meaningless fantasy. On the contrary, this story – including all the stuff about an 'I' which wills and wants and feels – is in its own terms a *valid* description of the mechanism which is causally responsible for human behaviour, namely the human brain. And I mean valid in the dictionary sense of 'fulfilling all the necessary conditions' for what a description of the human brain ought to be like, if such a description is to be used by human beings to understand the human behaviour which is produced by human brains.

This suggestion, admittedly, may seem a little puzzling. Quite apart from the oddity of suggesting that it is the *brain* which ordinary people are describing, when most people have never so much as seen a brain (and many of them do not even know they have one), there is a more obvious reason to be sceptical. For it may be thought that, *as* a description of a brain, the story of the 'I' simply does not ring true. Students of physiology have been studying brains for several hundred years, and the description they give is a quite different one. Brains, they tell us, are made of nerve-cells, chemicals and electricity. When a surgeon opens the head of a human being he does not find a soul.

No, he does not. Neither does a tourist who visits Oxford find the University. Nor did Pontius Pilate, when he cross-examined Jesus, find the Son of God. There are, as we shall see, different ways of describing the same thing, all of which may be valid in their way. To Pilate's famous question 'What is truth?', the answer must be 'It depends on who you are, and what you are trying to do.'

Suppose, for the sake of illustration, that we were to seek a valid description of that remarkable organ which constitutes the metaphorical nerve-centre of British politics, the House of Commons. Imagine two observers, with rather different interests, present on the same occasion in the Press Gallery: one is a maker of documentary films, a foreigner who knows nothing of the habits of the British legislature and is anxious to write down exactly what goes on; the other is the parliamentary correspondent of *The Times*.

The film-maker's description might read like this:

> Between 16.00 and 18.00 hours on Friday afternoon, men and women on each side of the chamber shouted at each other, threw bits of paper, clapped their hands, stamped their feet, and talked about cod; then, at a signal from an old fellow in a curly wig, everyone got to their feet, and those who had previously been sitting on the right side of the chamber all made their way out through a door marked A Y E, while those who had previously been sitting on the left side all made their way out through a door marked N O E.

The parliamentary correspondent's description like this:

> Following Prime Minister's Question Time, there ensued a lively debate on the White-Fish Fisheries Bill; the Speaker called for a vote, whereupon the House divided, and, under the influence of a three-line whip, the motion to give the Bill a second reading was carried for the Government.

Two very different descriptions of the same events, both of them accurate, both arguably valid; and yet with no obvious resemblance one to the other. Without doubt the film-maker's description would help him to reconstruct, if he later wanted to, a picture of what actually happened in the chamber of the House of Commons on that momentous afternoon. But it is surely the newspaper correspondent's description which would be more helpful to a parliamentary historian.

And yet the latter description, if it were to fall on unsympathetic ears, might well be dismissed as nonsensical guff. For it is a story which, on the face of it, is quite as illogical, ungrammatical and bordering on the metaphysical as anything the natural psychologist has to offer. Who ever heard of a 'house dividing', or of a 'motion being carried'? What is this thing, the 'Government'? Is it a thing at all? And yet . . . the correspondent's description does make the greater contribution to parliamentary history.

Let us return to the brain. The point of my example is not that brain physiologists bear any close resemblance to makers of documentary films (though a reader of, say, the *Journal of Neurophysiology* might sometimes innocently think it). There are, I am well aware, a good many physiologists who, for all their talk about chemicals and nerve-cells, do have a genuine interest in higher 'mental' processes – in the global achievements as well as the particular antics of the nerve-cells they describe. But my point is that just because the brain *can* be described in physiological terms, it does not mean it *ought* to be. Indeed the physiologist's description, so far from being the only valid description of the brain, is necessarily one among many alternative descriptions which may be more or less suited to a particular observer. It depends, as I said, on who the observer is and what he is trying to do. And if he is an ordinary human being, trying to understand behaviour, then the physiologist's description is likely to be almost worthless to him. Not even the specialist has yet been – and I doubt that he ever will be – able to make other than trivial predictions about human behaviour by direct reference to the fact that such behaviour is (as it is) ultimately controlled by nerve-cells, chemicals and electricity.

What the ordinary person needs in performing the workaday task of the natural psychologist is a description of the brain which, like that of the parliamentary correspondent, has been precisely crafted for the job in hand: and that means a description which is both relevant and accessible. Relevant to answering the questions which human beings daily ask about the inner causes of behaviour, and accessible to their wide but not unbounded powers of reasoning and imagination.

It is, I suggest, just such a description which reflexive consciousness provides. When the natural psychologist looks into himself and observes what he doth when he does think, hope, fear &c., he is enabled to read off (as it were from the columns of *The Times*) a relevant and accessible commentary on the parliamentary proceedings of his own brain: a commentary which is certainly highly selective, condensed, partial and presumptuous, but which none the less tells him most of what he needs to know, in a form which he is predisposed to understand.

2
THE SOCIAL FUNCTION OF INTELLECT

Henry Ford, it is said, commissioned a survey of the car scrap-yards of America to find out if there were parts of the Model T Ford which never failed. His inspectors came back with reports of almost every kind of failure: axles, brakes, pistons – all were liable to go wrong. But they drew attention to one notable exception, the *kingpins* of the scrapped cars invariably had years of life left in them. With ruthless logic Ford concluded that the kingpins on the Model T were too good for their job and ordered that in future they should be made to an inferior specification.

Nature is surely at least as careful an economist as Henry Ford. It is not her habit to tolerate needless extravagance in the animals on her production lines: superfluous capacity is trimmed back, new capacity added only as and when it is needed. We do not expect to find that animals possess abilities which far exceed the calls that natural living makes on them. If someone were to argue – as I shall suggest he might argue – that some primate species (and mankind in particular) are much cleverer than they need be, we know that he is most likely to be wrong. But it is not clear why he would be wrong. This chapter explores a possible answer. It is an answer which has meant for me a rethinking of the function of intellect.

A rethinking, or merely a first-thinking? I had not previously given much thought to the biological function of intellect, and my impression is that few others have done either. In the literature on animal intelligence there has been surprisingly little discussion of how intelligence contributes to biological fitness. Comparative psychologists have established that animals of one species perform better, for instance, on the Hebb–Williams maze than those of another, or that they are quicker to pick up learning sets or more successful on an 'insight' problem; there have been attempts to relate performance on particular kinds of tests to particular underlying cognitive skills; there has (more recently) been debate on how the same skill is to be assessed with 'fairness' in animals of different species; but there has seldom been consideration given to why the animal, in its natural environment, should *need* such skill. What is the use of 'conditional oddity discrimination' to a monkey in the field?[1] What advantage is there to an anthropoid ape in being able to recognise its own reflection in a mirror?[2] While it might indeed be 'odd for a biologist to make it his task to explain why horses can't learn mathematics',[3] it would not be odd for him to ask why *people can*.

The absence of discussion on these issues may reflect the view that there is little to discuss. It is tempting, certainly, to adopt a broad definition of intelligence which makes it self-evidently functional. Take, for instance, Alice Heim's definition of intelligence in man, 'the ability to grasp the essentials of a situation and respond appropriately':[4] substitute 'adaptively' for 'appropriately' and the problem of the biological function of intellect is (tautologically) solved. But even those definitions which are not so manifestly circular tend none the less to embody value-laden words. When intelligence is defined as the 'ability' to do this or that, who dares question the biological advantage of being *able*? When reference is made to 'understanding' or 'skill at problem-solving' the terms themselves seem to quiver with adaptiveness. Every animal's world is, after all, full of things to be understood and problems to be solved. For sure, the world is full of problems – but what exactly are these problems, how do they differ from animal to animal, and what particular advantage accrues to the individual who can solve them? These are not trivial questions.

Despite what has been said, we had better have a definition of intelligence, or the discussion is at risk of going adrift. The following formula provides at least some kind of anchor: 'An animal displays intelligence when it modifies its behaviour on the basis of valid inference from evidence.' The word 'valid' is meant to imply only that the inference is logically sound; it leaves open the question of how the animal benefits in consequence. This definition is admittedly wide, since it embraces everything from simple associative learning to syllogistic reasoning. Within the spectrum it seems fair to distinguish 'low-level' from 'high-level' intelligence. It requires, for instance, relatively low-level intelligence to infer that something is likely to happen merely because similar things have happened in comparable circumstances in the past; but it requires high-level intelligence to infer that something is likely to happen because it is entailed by a *novel* conjunction of events. The former is, I suspect, a comparatively elementary skill and widespread through the animal kingdom, but the latter is much more special, a mark of the 'creative' intellect which is characteristic especially of the higher primates. In what follows I shall be enquiring into the function chiefly of 'creative' intellect.

Now I am about to set up a straw man. But he is a man whose reflection I have seen in my own mirror, and I am inclined to treat him with respect. The opinion he holds is that the main role of creative intellect lies in *practical invention*. 'Invention' here is being used broadly to mean acts of intelligent discovery by which an animal comes up with new ways of doing things. Thus it includes not only, say, the fabrication of new tools or the putting of existing objects to new use but also the discovery of new behavioural strategies, new ways of using the resources of one's own body. But, wide as its scope may be, the talk is strictly of 'practical' invention, and in this context 'practical' has a restricted meaning. For the man in question sees the need for invention as arising only in relation to the external physical environment; he has not noticed – or has not thought it important – that many animals are *social* beings.

You will see, no doubt, that I have deliberately built my straw man with feet of clay. But let us none the less see where he stands. His idea of the intellectually challenging environment has been perfectly described by Daniel Defoe. It is the desert island of Robinson Crusoe – before the arrival of Man Friday. The island is a lonely, hostile environment, full of technological challenge, a world in which Crusoe depends for his survival on his skill in gathering food, finding shelter, conserving energy, avoiding danger. And he must work fast, in a truly inventive way, for he has no time to spare for learning simply by induction from experience. But was that the kind of world in which creative intellect evolved? I believe, for reasons I shall come to, that the real world was never like that, and yet that the real world of the higher primates may in fact be considerably *more* intellectually demanding. My view – and Defoe's, as I understand him – is that it was the arrival of Man Friday on the scene which really made things difficult for Crusoe. If Monday and Tuesday, Wednesday and Thursday had turned up as well then Crusoe would have had every need to keep his wits about him.

But the case for the importance of practical invention must be taken seriously. There can be no doubt that for some species in some contexts inventiveness does seem to have survival value. The 'subsistence technology' of chimpanzees[5] and even more that of 'natural' man[6] involves many tricks of technique which appear prima facie to be products of creative intellect. And what is true for these anthropoids must surely be true at least in part for other species. Animals who are quick to realise new techniques (in hunting, searching, navigating or whatever) would seem bound to gain in terms of fitness. Why, then, should one dispute that there have been selective pressures operating to bring about the evolution of intelligence in relation to practical affairs? I do not of course dispute the general principle; what I question is how much this principle *alone* explains. How clever does a man or monkey need to be before the returns on superior intellect become vanishingly small? If, despite appearances, the important practical problems of living actually demand only relatively low-level intelligence for their solution, then there would be grounds for supposing that high-level creative intelligence is wasted. Even Einstein could not get better than 100 per cent at O level. Can we really explain the evolution of the higher intellectual faculties of primates on the basis of success or failure in their 'practical exams'?

My answer is no, for the following reason: even in those species which have the most advanced technologies the exams are largely tests of knowledge rather than of imaginative reasoning. The evidence from field studies of chimpanzees all points to the fact that subsistence techniques are hardly if ever the product of premeditated invention; they are arrived at instead either by trial-and-error learning or by imitation of others. Indeed it is hard to imagine how many of the techniques could in principle be arrived at otherwise. G. Teleki concluded on the basis of his own attempts at 'termiting' that there was no way of predicting a priori what would be the most effective kind of probe to stick into a termite hill, or how best to twiddle it or, for that matter, where to stick it.[7] He had to learn inductively by trial and error or, better, by mimicking the behaviour of Leakey, an old and experienced chimpanzee.Thus the chimpanzees' art would seem to be no more an invention than is the uncapping of milk-bottles by tits. And even where a technique could in principle be invented by deductive reasoning there are generally no grounds for supposing that it has been. Termiting by human beings is a case in point. In northern Zaire, people beat with sticks on the top of termite mounds to encourage the termites to come to the surface. The technique works because the stick-beating makes a noise like falling rain. It is just possible that someone once upon a time noticed the effect of falling rain, noticed the resemblance between the sound of rain and the beating of sticks, and put two and two together. But I doubt if that is how it happened; serendipity seems a much more likely explanation. Moreover, whatever the origin of the technique, there is certainly no reason to invoke inventiveness on the part of present-day practitioners, for these days stick-beating is culturally transmitted. My guess is that most of the practical problems that face higher primates can, as in the case of termiting, be dealt with by learned strategies without recourse to creative intelligence.

Paradoxically, I would suggest that subsistence technology, rather than requiring intelligence, may actually become a substitute for it. Provided the *social* structure of the species is such as to allow individuals to acquire subsistence techniques by simple associative learning, then there is little need for individual creativity. Thus the chimpanzees at Gombe, with their superior technological culture, may in fact have *less* need than the neighbouring baboons to be individually inventive. Indeed there might seem on the face of it to be a negative correlation between the intellectual capacity of a species and the need for intellectual output. The great apes, demonstrably the most intellectually gifted of all non-human animals, seem on the whole to lead comparatively undemanding lives, less demanding than those not only of lower primates but also of many non-primate species. During two months I spent watching gorillas in the Virunga mountains of Rwanda I could not help being struck by the fact that of all the animals in the forest the gorillas seemed to lead much the simplest existence – food abundant and easy to harvest (provided they *knew* where to find it), few if any predators (provided they *knew* how to avoid them) . . . little to do in fact (and little done) but eat, sleep and play. And the same is arguably true for natural man. Studies of contemporary Bushmen suggest that the life of hunting and gathering, typical of early man, was probably a remarkably easy one. The 'affluent savage'[8] seems to have established a *modus vivendi* in which, for a period of perhaps five million years, he could afford to be not only physically but intellectually lazy.

We are thus faced with a conundrum. It has been repeatedly demonstrated in the artificial situations of the psychological laboratory that anthropoid apes possess impressive powers of creative reasoning, yet these feats of intelligence seem simply not to have any parallels in the behaviour of the same animals in their natural environment. I have yet to hear of any example from the field of a chimpanzee (or for that matter a Bushman) using his full capacity for inferential reasoning in the solution of a biologically relevant practical problem. Someone may retort that if an ethologist had kept watch on Einstein through a pair of field-glasses he might well have come to the conclusion that Einstein too had a humdrum mind. But that is just the point: Einstein, like the chimpanzees, displayed his genius at rare times in 'artificial' situations – he did not use it, for he did not *need* to use it, in the common world of practical affairs.

Why then do the higher primates need to be as clever as they are and, in particular, that much cleverer than other species? What – if it exists – is the natural equivalent of the laboratory test of intelligence? The answer has, I believe, been ripening on the tree of the preceding discussion. I have suggested that the life of the great apes and man may not require much in the way of practical invention, but it does depend critically on the possession of wide factual knowledge of practical technique and the nature of the habitat. Such knowledge can only be acquired in the context of a *social* community – a community which provides both a medium for the cultural transmission of information and a protective environment in which individual learning can occur. I propose that the chief role of creative intellect is to hold society together.

In what follows I shall try to explain this proposal, to justify it, and to examine some of its surprising implications.

To me, as a Cambridge-taught psychologist, the proposal is in fact a rather strange one. Experimental psychologists in Britain have tended to regard social psychology as a poor country cousin of their subject – gauche, undisciplined and slightly absurd. Let me recount how I came to a different way of thinking, since this personal history will lead directly in to what I want to say. Some years ago I made a discovery which brought home to me dramatically the fact that, even for an experimental psychologist, a *cage* is a bad place in which to keep a monkey. I was studying the recovery of vision in a rhesus monkey, Helen, from whom the visual cortex had been surgically removed.[9] In the first four years I'd worked with her Helen had regained a considerable amount of visually guided behaviour, but she still showed no sign whatever of three-dimensional spatial vision. During all this time she had, however, been kept within the confines of a small laboratory cage. When, at length, five years after the operation, she was released from her cage and taken for walks in the open field at Madingley her sight suddenly burgeoned and within a few weeks she had recovered almost perfect spatial vision. The limits on her recovery had been imposed directly by the limited environment in which she had been living. Since that time, in working with laboratory monkeys I have been mindful of the possible damage that may have been done to them by their impoverished living conditions. I have looked anxiously through the wire mesh of the cages at Madingley, not only at my own monkeys but at Robert Hinde's. Now, Hinde's monkeys are rather better off than mine. They live in social groups of eight or nine animals in relatively large cages. But these cages are almost empty of objects, there is nothing to manipulate, nothing to explore; once a day the concrete floor is hosed down, food pellets are thrown in and that is about it. So I looked – and seeing this barren environment, thought of the stultifying effect it must have on the monkey's intellect. And then one day I looked again and saw a half-weaned infant pestering its mother, two adolescents engaged in a mock battle, an old male grooming a female while another female tried to sidle up to him, and I suddenly saw the scene with new eyes: forget about the absence of *objects*, these monkeys had *each other* to manipulate and to explore. There could be no risk of their dying an intellectual death when the social environment provided such obvious opportunity for participating in a running dialectical debate. Compared to the solitary existence of my own monkeys, the set-up in Hinde's social groups came close to resembling a simian School of Athens.

The scientific study of social interaction is not far advanced. Much of the best published literature is in fact genuinely 'literature' – Aesop and Dickens make, in their own way, as important contributions as Laing, Goffman or Argyle. But one generalisation can I think be made with certainty: the life of social animals is highly problematical. In a complex society, such as those we know exist among higher primates, there are benefits to be gained for each individual member both from preserving the overall structure of the group and at the same time from exploiting and out-manoeuvring others within it. Thus social primates are required by the very nature of the system they create and maintain to be calculating beings; they must be able to calculate the consequences of their own behaviour, to calculate the likely behaviour of others, to calculate the balance of advantage and loss – and all this in a context where the evidence on which their calculations are based is ephemeral, ambiguous and liable to change, not least as a consequence of their own actions. In such a situation, 'social skill' goes hand in hand with intellect, and here at last the intellectual faculties required are of the highest order. The game of social plot and counter-plot cannot be played merely on the basis of accumulated knowledge, any more than can a game of chess.

Like chess, a social interaction is typically a *trans*action between social partners. One animal may, for instance, wish by his own behaviour to change the behaviour of another; but since the second animal is himself reactive and intelligent the interaction soon becomes a two-way argument where each 'player' must be ready to change his tactics – and maybe his goals – as the game proceeds. Thus, over and above the cognitive skills which are required merely to perceive the current state of play (and they may be considerable), the social gamesman, like the chess-player, must be capable of a special sort of forward planning. Given that each move in the game may call forth several alternative responses from the other player this forward planning will take the form of a decision tree, having its root in the current situation and branches corresponding to the moves considered in looking ahead at different possibilities. It asks for a level of intelligence which is, I submit, unparalleled in any other sphere of living. There may be, of course, strong and weak players – yet, as master or novice, we and most other members of complex primate societies have been in this game since we were babies.

But what makes a society 'complex' in the first place? There have probably been selective pressures of two rather different kinds, one from without, the other from within society. I suggested that one of the chief functions of society is to act as it were as a polytechnic school for the teaching of subsistence technology. The social system serves the purpose in two ways: (i) by allowing a period of prolonged dependence during which young animals, spared the need to fend for themselves, are free to experiment and explore; and (ii) by bringing the young into contact with older, more experienced members of the community from whom they can learn by imitation (and perhaps, in some cases, from more formal 'lessons'). Now, to the extent that this kind of education has adaptive consequences, there will be selective pressures both to prolong the period of untrammelled infantile dependency (to increase the 'school-leaving age') and to retain older animals within the community (to increase the number of experienced 'teachers'). But the resulting mix of old and young, caretakers and dependants, sisters, cousins, aunts and grandparents not only calls for considerable social responsibility but also has potentially disruptive social consequences. The presence of dependants (young, injured or infirm) clearly calls at all times for a measure of tolerance and unselfish sharing. But in so far as biologically important resources may be scarce (as subsistence materials must sometimes be, and sexual partners will be commonly) there is a limit to which tolerance can go. Squabbles are bound to occur about access to these scarce resources and different individuals will have different interests in participating in, promoting or putting a stop to such squabbles. Thus the stage is set within the 'collegiate community' for considerable political strife. To do well for oneself while remaining within the terms of the social contract on which the fitness of the whole community ultimately depends calls for remarkable reasonableness (in both literal and colloquial senses of the word). It is no accident therefore that men, who of all primates show the longest period of dependence (nearly thirty years in the case of Bushmen), the most complex kinship structures, and the widest overlap of generations within society, should be more intelligent than chimpanzees, and chimpanzees for the same reasons more intelligent than monkeys.

Once a society has reached a certain level of complexity, then new internal pressures must arise which act to increase its complexity still further. For, in a society of the kind outlined, an animal's intellectual 'adversaries' are members of his own breeding community. If intellectual prowess is correlated with social success, and if social success means high biological fitness, then any heritable trait which increases the ability of an individual to outwit his fellows will soon spread through the gene pool. And in these circumstances there can be no going back: an evolutionary 'ratchet' has been set up, acting like a self-winding watch to increase the general intellectual standing of the species. In principle the process might be expected to continue until either the physiological mainspring of intelligence is full-wound or else intelligence itself becomes a burden. The latter seems most likely to be the limiting factor: there must surely come a point where the time required to resolve a social 'argument' becomes insupportable.

The question of the time given up to unproductive social activity is an important one. Members of the community – even if they have not evolved a runaway intellect – are bound to spend a considerable part of their lives in caretaking and social politics. It follows that they inevitably have less time to spare for basic subsistence activities. If the social system is to be of any net biological benefit the improvement in subsistence techniques which it makes possible must more than compensate for the lost time. To put the matter baldly: if an animal spends all morning in non-productive socialising, he must be at least twice as efficient a producer in the afternoon. We might therefore expect that the evolution of a social system capable of supporting advanced technology should only happen under conditions where improvements in technique can substantially increase the return on labour. This may not always be the case. To take an extreme example, the open sea is probably an environment where technical knowledge can bring little benefit, so that complex societies – and high intelligence – are contra-indicated (dolphins and whales provide, maybe, a remarkable and unexplained exception). Even at Gombe the net advantage of having a complex social system may in fact be marginal; the chimpanzees at Gombe share several of the local food resources with baboons, and it would be instructive to know how far the advantage that chimpanzees have over baboons in terms of technical skill is eroded by the relatively large amount of time they give up to social intercourse. It may be that what the chimpanzees gain on the swings of technical proficiency they lose on the roundabouts of extravagant socialising. As it is, in a year of poor harvest the chimpanzees in fact become much less sociable;[10] my guess is that they simply cannot spare the time. The ancestors of man, however, when they moved into the savanna, discovered an environment where technical knowledge began to pay new and continuing dividends. It was in that environment that the pressures to give children an even better schooling created a social system of unprecedented complexity – and with it unprecedented challenge to intelligence.

The outcome has been the gifting of members of the human species with remarkable powers of social foresight and understanding. This social intelligence, developed initially to cope with local problems of interpersonal relationships, has in time found expression in the institutional creations of the 'savage mind' – the highly rational structures of kinship, totemism, myth and religion which characterise primitive societies.[11] And it is, I believe, essentially the same intelligence which has created the systems of philosophical and scientific thought which have flowered in advanced civilisations in the last four thousand years. Yet civilisation has been too short-lived to have had any important evolutionary consequences; the 'environment of adaptiveness'[12] of human intelligence remains the *social* milieu.

If man's intellect is thus suited primarily to thinking about people and their institutions, how does it fare with *non-social* problems? To end this chapter I want to raise the question of 'constraints' on human reasoning, such as might result if there is a predisposition to try to fit non-social material into a social mould.

When a person sets out to solve a social problem he may reasonably have certain expectations about what he is getting in to. First, he should know that the situation confronting him is unlikely to remain stable. Any social transaction is by its nature a developing process and the development is bound to have a degree of indeterminacy to it. Neither of the social agents involved in the transaction can be certain of the future behaviour of the other; as in Alice's game of croquet with the Queen of Hearts, both balls and hoops are always on the move. Someone embarking on such a transaction must therefore be prepared for the problem itself to alter as a consequence of his attempt to solve it – in the very act of interpreting the social world he changes it. Like Alice he may well be tempted to complain 'You've no idea how confusing it is, all the things being alive'; that is not the way the game is played at Hurlingham – and that is not the way that non-social material typically behaves. But, secondly, he should know that the development *will* have a certain logic to it. In Alice's croquet game there was real confusion, everyone played at once without waiting for turns, and there were no rules; but in a social transaction there are, if not strict rules, at least definite constraints on what is allowed and definite conventions about how a particular action by one of the transactors should be answered by the other. My earlier analogy with the chess game was perhaps a more appropriate one; in social behaviour there is a kind of turn-taking, there are limits on what actions are allowable, and at least in some circumstances there are conventional, often highly elaborated, sequences of exchange.

Even the chess analogy, however, misses a crucial feature of social interaction. For while the good chess player is essentially selfish, playing only to win, the selfishness of social animals is typically tempered by what, for want of a better term, I would call *sympathy*. By sympathy I mean a tendency on the part of one social partner to identify himself with the other and so to make the other's goals to some extent his own. The role of sympathy in the biology of social relationships has yet to be thought through in detail, but it is probable that sympathy and the 'morality' which stems from it[13] are biologically adaptive features of the social behaviour of both men and other animals – and consequently major constraints on 'social thinking' wherever it is applied. Thus our man setting out to apply his intelligence to solve a social problem may expect to be involved in a fluid, transactional exchange with a sympathetic human partner. To the extent that the thinking appropriate to such a situation represents the customary mode of human thought, men may be expected to behave inappropriately in contexts where a transaction cannot in principle take place: if they treat inanimate entities as 'people' they are sure to make mistakes.

There are many examples of fallacious reasoning which would fit such an interpretation. The most obvious cases are those where men do in fact openly resort to animistic thinking about natural phenomena. Thus primitive – and not so primitive – peoples commonly attempt to *bargain* with nature, through prayer, through sacrifice or through ritual persuasion. In doing so they are explicitly adopting a social model, expecting nature to participate in a transaction. But nature will not transact with men; she goes her own way regardless – while her would-be interlocutors feel grateful or slighted as befits the case. Transactional thinking may not always be so openly acknowledged, but it often lies just below the surface in other cases of 'illogical' behaviour. Thus the gambler at the roulette table, who continues to bet on the red square precisely because he has already lost on red repeatedly, is behaving as though he expects the behaviour of the roulette wheel to respond eventually to his persistent overtures; he does not – as he would be wise to do – conclude that the odds are unalterably set against him. Likewise, the man in P. C. Wason's experiments on abstract reasoning who, when he is given the task of discovering a mathematical rule, typically tries to substitute *his own* rule for the predetermined one is acting as though he expects the problem itself to change in response to his trial solutions.[14] The comment of one of Wason's subjects is revealing: 'Rules are relative. If you were the subject, and I were the experimenter, then I would be right.' In general, I would suggest, a transactional approach leads men to refuse to accept the intransigence of facts – whether the facts are physical events, mathematical axioms or scientific laws; there will always be the temptation to assume that the facts will respond like living beings to social pressures. Men expect to argue *with* problems rather than being limited to arguing *about* them.

There are times, however, when such a 'mistaken' approach to natural phenomena can be unexpectedly creative. While it may be the case that no amount of social pleading will change the weather or, for that matter, transmute base metals into gold, there are things in nature with which a kind of social intercourse is possible. It is not strictly true that nature will not transact with men. If we mean by a transaction essentially a developing relationship founded on mutual give and take, then several of the relationships which men enter into with the non-human things around them may be considered to have transactional qualities. The cultivation of plants provides a clear and interesting example: the care which a gardener gives to his plants (watering, fertilising, hoeing, pruning etc.) is attuned to the plants' emerging properties, which properties are in turn a function of the gardener's behaviour. True, plants will not respond to ordinary social pressures (though men *do* talk to them), but the way in which they give to and receive from a gardener bears, I suggest, a close structural similarity to a simple social relationship. If C. Trevarthen can speak of 'conversations' between a mother and a two-month-old baby,[15] so too might we speak of a conversation between a gardener and his roses or a farmer and his corn. And the same can be argued for men's interactions with certain wholly inanimate materials. The relationship of a potter to his clay, a smelter to his ore or a cook to his soup are all relationships of fluid mutual exchange, again proto-social in character.

It is not just that transactional thinking is typical of man; transactions are something which people actively seek out and will force on nature wherever they are able. In the Doll Museum in Edinburgh there is a case full of bones clothed in scraps of rag – moving reminders of the desire of human children to conjure up social relationships with even the most unpromising material. Through a long history, men have, I believe, explored the transactional possibilities of countless of the things in their environment, and sometimes, Pygmalion-like, the things have come alive. Thus many of mankind's most prized technological discoveries, from agriculture to chemistry, may have had their origin not in the deliberate application of practical intelligence but in the fortunate misapplication of social intelligence.

The rise of classical scientific method has in large measure depended on human thinkers disciplining themselves to abjure transactional, socio-magical styles of reasoning. But scientific method has come to the fore only in the last few hundred years of mankind's history, and in our own times there are everywhere signs of a return to more magical systems of interpretation. In dealing with the non-social world the former method is undoubtedly the more immediately appropriate; but the latter is perhaps more natural to man. Transactional thinking may indeed be irrepressible: within the most disciplined Jekyll is concealed a transactional Hyde. Charles Dodgson the mathematician shared his pen amicably enough with Lewis Carroll the inventor of Wonderland, but the split is often neither so comfortable nor so complete. Newton is revealed in his private papers as a Rosicrucian mystic, and his intellectual descendants continue to this day to apply strange double standards to their thinking – witness the way in which certain British physicists took up the cause of Uri Geller, the man who, by wishing it, could bend a metal spoon.[16] In the long view of science, there is, I suspect, good reason to approve this kind of inconsistency. For while 'normal science' (in Kuhn's sense of the term) has little if any room for social thinking, 'revolutionary science' may more often than we realise derive its inspiration from a vision of a socially transacting universe. Particle physics has already followed Alice down the rabbit-hole into a world peopled by 'families' of elementary particles endowed with 'strangeness' and 'charm'. *Vide*, for example, the following report in *New Scientist*: 'The particles searched for at SPEAR were the *cousins* of the psis made from one *charm* quark and one *uncharmed* antiquark. This contrasts with the *siblings* of the psis . . .' (my italics).[17] Who knows where such 'socio-physics' may eventually lead?

The ideology of classical science has had a huge but in many ways narrowing influence on ideas about the nature of 'intelligent' behaviour. But no matter what the high priests, from Bacon to Popper, have had to say about how people ought to think, they have never come near to describing how people *do* think. In so far as an idealised view of scientific method has been the dominant influence on mankind's recent intellectual history, biologists should be the first to follow Henry Ford in dismissing recent history as 'bunk'. Evolutionary history, however, is a different matter. The formative years for human intellect were the years when man lived as a social savage on the plains of Africa. Even now, as Sir Thomas Browne wrote in *Religio medici*, 'All Africa and her prodigies are within us.'

3

NATURE'S PSYCHOLOGISTS

In *The Nature of Explanation* Kenneth Craik outlined a 'hypothesis on the nature of thought', proposing that

> the nervous system is . . . a calculating machine capable of modelling or paralleling external events . . . If the organism carries a 'small-scale model' of external reality and of its own possible actions within its head, it is able to try out various alternatives, conclude which is the best of them, react to future situations before they arise, utilise the knowledge of past events in dealing with the future, and in every way to react in a much fuller, safer and more competent manner to the emergencies which face it.[1]

The notion of a 'mental model of reality' has become in the years since so widely accepted that it has grown to be almost a cliché of experimental psychology. And like other clichés its meaning is no longer called in question. From the outset Craik's 'hypothesis' begged some fundamental questions: A model of *reality*? What reality? Whose reality?

My dog and I live in the same house. Do we share the same 'reality'? Certainly we share the same physical environment, and most aspects of that physical environment are probably as real for one of us as for the other. Maybe our realities differ only in the trivial sense that we each know a few things about the house that the other does not – the dog (having a better nose than I) knows better the smell of the carpet, I (having a better pair of eyes) know better the colour of the curtain. Now, suppose my dog chews up the gas bill which is lying on the mat by the door. Is the reality of that event the same for him as me? Something real enough has happened for us both, and the same piece of paper is involved. The dog hangs his head in contrition. Is he contrite because he has chewed up the gas bill? What does a dog know about *gas bills*! Gas bills are an important part of my external reality, but they are surely none of his.

If my and the dog's realities differ in this and other more important ways they do so because we have learned to conceptualise the world on different lines. To the dog paper is paper, to me it is newspaper, wrapping paper or a letter from my friend. These ways of looking at paper are essentially human ways, conditioned of course by culture, but a culture which is a product of a specifically human nature. I and the dog are involved with different aspects of reality because, at bottom, we are *biologically adapted* to lead different kinds of lives.

To all biological intents and purposes the portion of reality which matters to any particular animal is that portion of which it must have a working knowledge in the interests of its own survival. Because animals differ in their life-styles they face different kinds of 'emergencies' and they must therefore have different kinds of knowledge if they are to react in the full, safe, competent manner which Craik – and natural selection – recommends.

But different kinds of knowledge entail different ways of knowing. In so far as animals are biologically adapted to deal specifically with their own portions of reality, so must their nervous 'calculating machines' be adapted to construct very different kinds of models. That is not to say merely that the calculating machines may be required to do different kinds of sums, but rather that they may have to work according to quite different heuristic principles. Depending on the job for which nature has designed them the nervous systems will differ in the kind of *concepts* they employ, the *logical calculus* they use, the *laws of causation* they assume, and so on.

Let us call these nervous calculating machines 'minds'. It is the thesis of this chapter that a revolutionary advance in the evolution of mind occurred when, for certain social animals, a new set of heuristic principles was devised to cope with the pressing need to model a special section of reality – the reality comprised by the behaviour of other kindred animals. The trick which nature came up with was *introspection*; it proved possible for an individual to develop a model of the behaviour of others by reasoning by analogy from his own case, the facts of his own case being revealed to him through examination of the contents of consciousness.

For man and other animals which live in complex social groups reality is in larger measure a 'social reality'. No other class of environmental objects approaches in biological significance those living bodies which constitute for a social animal its companions, playmates, rivals, teachers, foes. It depends on the bodies of other conspecific animals not merely for its immediate sustenance in infancy and its sexual fulfilment as an adult, but in one way or another for the success (or failure) of almost every enterprise it undertakes. In these circumstances the ability to model the behaviour of others in the social group has paramount survival value.

I have argued in more detail before now that the modelling of other animals' behaviour is not only the most important but also the most difficult task to which social animals must turn their minds (see Chapter 2). In retrospect I do not think I took my own case seriously enough. The task of modelling behaviour does indeed demand formidable intellectual skill – social animals have evolved for that reason to be the most intelligent of animals – but intelligence alone is not enough. If a social animal is to become – as it must – one of 'nature's psychologists' it must somehow come up with the appropriate framework for doing psychology; it must develop a fitting set of concepts and a fitting logic for dealing with a unique and uniquely elusive portion of reality.

The difficulties that arise from working within an *inappropriate* framework are well enough illustrated by the history of the science of experimental psychology. For upwards of a hundred years academic psychologists have been attempting, by the 'objective' methods of the physical sciences, to acquire precisely the kind of knowledge of behaviour which every social animal must have in order to survive. In so far as these psychologists have been strict 'behaviourists' they have gone about their task as if they were studying the behaviour of billiard-balls, basing their theoretical models entirely on concepts to which they could easily give public definition. And in so far as they have been strict behaviourists they have made slow progress. They have been held up again and again by their failure to develop a sufficiently rich or relevant framework of ideas. Concepts such as 'habit strength', 'drive' or 'reinforcement', for all their objectivity, are hopelessly inadequate to the task of modelling the subtleties of real behaviour. Indeed, I venture to suggest that if a rat's knowledge of the behaviour of other rats were to be limited to everything which behaviourists have discovered about rats to date, the rat would show so little understanding of its fellows that it would bungle disastrously every social interaction it engaged in; the prospects for a man similarly constrained would be still more dismal. And yet, as professional scientists, behaviourists have always had enormous advantages over an individual animal, being able to do controlled experiments, to subject their data to sophisticated statistical analysis, and above all to share the knowledge recorded in the scientific literature. By contrast, an animal in nature has only its own experience to go on, its own memory to record it, and its own brief lifetime to acquire it. 'Behaviourism' as a philosophy for the *natural* science of psychology could not, and presumably does not, fit the bill.

Noam Chomsky in his famous review of B. F. Skinner's *Verbal Behaviour* argued on parallel lines that it would be impossible for a child to acquire an understanding of human spoken language if all the child had at its disposal was a clever brain with which to make an unprejudiced analysis of public utterances.[2] Chomsky's way round the problem was to propose that the child's brain is not in fact unprejudiced: the child is born with an innate knowledge of transformational grammar, which provides it with the framework for modelling human language. Though there are snags about Chomsky's thesis, it would not, I suppose, be wholly unreasonable to suggest something similar with regard to the acquisition of a model of behaviour: the essential rules and concepts for understanding behaviour might simply be innately given to a social animal. There is, however, an alternative, more attractive possibility. This is to suggest that the animal has access not to 'innate knowledge' but to 'inside evidence' about behaviour. Nature's psychologists succeed where academic psychologists have failed because the former make free use of introspection.

Let us consider how introspection works. I shall write these paragraphs from the position of a reflective conscious human being, on the assumption that other human beings will understand me. First let me distinguish two separate meanings of what may be called 'self-observation', a weak one and a strong one. In the weak sense self-observation means simply observing my own body as opposed to someone else's. It is bound to be true that my body is the example of a human body which is far the most familiar to me. Thus even if I could only observe my behaviour through 'objective' eyes it is likely that I would draw on self-observation for most of my evidence about how a human being behaves (in the same way that a physicist who carried a billiard-ball about in his pocket might well use that 'personal' billiard-ball as the paradigm of billiard-balls in general).

But the importance of self-observation does not stop there. In the strong sense of the term self-observation means a special sort of observation to which I and I alone am privileged. When I reflect on my own behaviour I become aware not only of the external facts about my actions but of a conscious presence, 'I', which 'wills' those actions. This 'I' has *reasons* for the things it wills. The reasons are various kinds of 'feeling' – 'sensations', 'emotions', 'memories', 'desires'. 'I' want to eat because 'I' am hungry, 'I' intend to go to bed because 'I' am tired, 'I' refuse to move because 'I' am in pain. Moreover, experience tells me that the feelings themselves are caused by certain things which happen to my body in the outside world. 'I' am hungry because my body has been without food, 'I' am in pain because my foot has trodden on a thorn. It so happens (as I soon discover) that several sorts of happening may cause a particular feeling and that a particular feeling may be responsible for my willing several sorts of action. The role of a feeling in the model I develop of my own behaviour becomes, therefore, that of what psychologists have called an 'intervening variable', bridging the causal gap between a set of antecedent circumstances and a set of subsequent actions – between what happens to 'me' and what 'I' do.

Now, when I come to the task of modelling the behaviour of another man, I naturally assume that he operates on the same principles as I do. I assume that within him too there is a conscious 'I' and that his 'I' has feelings which are the reasons for 'his' willing certain actions. In other words I expect the relation between what happens to his body and what he does to have the same causal structure – a structure based on the same intervening variables – as I have discovered for myself. It is my familiarity with this causal structure and these variables which provides me with the all-important ideological framework for doing natural psychology.

Without introspection to guide me, the task of deciphering the behaviour of my fellow men would be quite beyond my powers. I should be like a poor cryptographer attempting to decipher a text which was written in a totally unfamiliar language. Michael Ventris could crack the code of Linear B because he guessed in advance that the language of the text was Greek; although the alphabet was strange to him he reckoned – correctly – that he knew the syntax and vocabulary of the underlying message. Linear A remains to this day a mystery because no one knows what language it is written in.

In so far as we are conscious human beings we all guess in advance the 'language' of other men's behaviour.

But it may be objected that I have not really made out a case for there being any unique advantage in using introspection, since non-introspective psychological scientists do in fact also allow themselves to postulate certain intervening variables such as 'hunger' and 'fear'. And so they do. But think of how they derive them. To establish what variables are likely to prove useful to their models they must (assuming they do not cheat) make a vast and impartial survey of all the circumstances and all the actions of an animal and then subject their data to statistical factor analysis. In practice, of course, they usually do cheat by restricting their data to a few 'relevant' parameters – relevance being decided on the basis of an *intuitive* guess. But even so their task is not an easy one. Before postulating even such an 'obvious' variable as hunger the experimental psychologist must go through a formidable exercise in data collection and statistical cross-correlation.[3] An ordinary introspective human being has, however, no such problem in devising a 'psychological' model of his own and other men's behaviour: he knows from his own internal feelings what intervening variables to go for. Indeed he knows of subtle feelings which no amount of objective data-crunching is likely to reveal as useful postulates. Speaking again for myself, I know of feelings of awe, of guilt, of jealousy, of irritation, of hope, of being in love, all of which have a place in my model of how other men behave.

Before I can attribute such feelings to others I must, it seems, myself have had them – a proviso which the academic psychologist is spared. But it is generally the case, for reasons I shall bring out in Chapters 6–8, that in the course of their lives most people do have most of them, and often indeed it takes only a single seminal experience to add a new dimension to one's behavioural model. Let a celibate monk just once make love to a woman and he would be surprised how much better he would understand the Song of Solomon; but let him, like an academic psychologist, observe twenty couples in the park and he would not be that much wiser . . .

So much for how I think that nature's psychologists proceed. Let me turn to the more purely philosophical implications of the theory, and say something further about *consciousness*.

I take it to be the case that what we mean by someone's conscious experience is the set of subjective feelings which, at any one time, are available to introspection: that is, the sensations, emotions, volitions etc. that I have talked of. Our criterion for judging that someone else is conscious is that we should have grounds for believing that he has subjective *reasons* for his actions – that he is eating an apple because he *feels hungry*, or that he is raising his arm because he *wants* to. If we had grounds for believing that a dog had similar subjective reasons for its actions we should want to say the dog was conscious too. In proposing a theory about the biological function of introspection I am therefore proposing a theory about the biological function of consciousness. And the implications of this theory are by no means trivial. If consciousness has evolved as a biological adaptation for doing introspective psychology, then the presence or absence of consciousness in animals of different species will depend on whether or not they need to be able to understand the behaviour of other animals in a social group. Wolves and chimpanzees and elephants, which all go in for complex social interactions, are probably all conscious; frogs and snails and codfish are probably not.

There may be philosophers who protest that it is nonsense to talk of a biological 'function' for consciousness when, so Wittgenstein tells us, conscious experience does not even have a 'place in the language game'.[4] But what Wittgenstein demonstrated is that there are logical problems about the *communication* of conscious experience – and it is not proposed by the theory that consciousness has any direct role in communication between individuals; I am not saying that social animals either can or should *report* their subjective feelings to each other. The advantage to an animal of being conscious lies in the purely private use it makes of conscious experience as a means of developing a conceptual framework which helps it to model another animal's behaviour. It need make no difference at all whether the other animal is actually experiencing the feelings with which it is being credited; all that matters is that its behaviour should be understandable on the assumption that such feelings provide the reasons for its actions. Thus for all I know no man other than myself has ever experienced a feeling corresponding to my own feeling of hunger; the fact remains that the concept of hunger, derived from my own experience, helps me to understand other men's eating behaviour. Indeed, if we assume that the first animal in history to have any sort of introspective consciousness occurred as a chance variant in an otherwise unconscious population, the selective advantage which consciousness gave that animal *must* have been independent of consciousness in others. It follows, *a fortiori*, that the selective advantage of consciousness can never have depended on one animal's conscious experience being the 'same' as another's.[5]

Maybe this sounds paradoxical. Indeed, if it does not sound a little paradoxical I should be worried. For I assume that you yourself are as naturally inclined as any other introspective animal to project your conscious feelings on to others. The suggestion that you may be wrong to do so, or at least that it does not matter whether you are right or wrong, does I hope arouse a certain Adamite resistance in you. But allow me to elaborate the argument.

I think no one would object to the claim that a piece of magnetised iron lacks consciousness. Suppose now that an animal – let us call it one of 'Nature's physicists' – wanted to model the behaviour of magnets. I can conceive that it might be helpful to that animal to think of the north pole of a magnet as having a *desire* to approach a south pole. Then, if the concept of having a desire was one which the animal knew about from its own inner experience, I should want to argue that introspective consciousness was an aid to the animal in doing physics. The fact that the animal would almost certainly be *incorrect* in attributing feelings of desire to magnets would be irrelevant to whether or not the attribution was heuristically helpful to it in developing a conceptual model of how magnets behave. But if this is conceivably true of doing physics, all the more is it true of doing psychology. Notwithstanding the logical possibility that every other human being around me is as unconscious as a piece of iron, my attribution of conscious feelings to them does as a matter of fact help me sort out my observations of their behaviour and develop predictive models.

Ah, you may say, but you are not really saying anything very interesting, since it can only be helpful to attribute feelings to other people – or magnets – in so far as there is *something* about the other person or the magnet which corresponds to what you call a feeling: the attribution of desire to magnets is heuristically valuable if, and only if, there exists in reality an electromagnetic attractive force between a north pole and a south pole, and the attribution of a feeling of hunger to a man is valuable if, and only if, his body is in reality motivated by a particular physiological state. Quite so. But the magnet does not have to know about the electromagnetic force and the man does not, in principle, have to know about the physiological state.

Magnets do not need to do physics. If they did – if their survival as magnets depended on it – perhaps they would be conscious. If volcanoes needed to do geology, and clouds needed to do meteorology, perhaps they would be conscious too.

But the survival of human beings *does* depend on their being able to do psychology. That is why, despite the sophistical doubts I have just expressed, I do not consider it to be even a biological possibility – let alone do I really believe – that other people are not as fully conscious of the reasons for their actions as I know that I myself am. In the case of frogs and snails and cod, however, my argument leads me to the opposite conclusion. Let me say it again: these non-social animals no more need to do psychology than magnets need to do physics – *ergo* they could have no use for consciousness.

Somewhere along the evolutionary path which led from fish to chimpanzees a change occurred in the nervous system which transformed an animal which simply 'behaved' into an animal which at the same time informed its mind of the reasons for its behaviour. My guess is that this change involved the evolution of a new brain – a 'conscious brain' parallel to the older 'executive brain'. In the last few years evidence has at last begun to emerge from studies of brain damage in animals and man which makes this kind of speculation meaningful.

To end this chapter I want to say more about the monkey, Helen, to whom I referred in Chapter 2.

In 1966 Helen underwent an operation on her brain in which the visual cortex was almost completely removed. In the months immediately following the operation she acted as if she were blind. But I and Professor Larry Weiskrantz, with whom I was working, were not convinced that Helen's blindness was so deep and permanent as it appeared. Could it be that her blindness lay not so much in her brain as in her mind? Was her problem that she did not *think* that she could see?

I set to work to persuade her to use her eyes again. Over the course of seven years I coaxed her, played with her, took her for walks – encouraged her in every way I could to realise her latent potential for vision. And slowly, haltingly, she found her way back from the dark valley into which the operation had plunged her. After seven years her recovery seemed so complete that an innocent observer would have noticed very little wrong with the way she analysed the visual world. She could, for example, run around a room full of furniture picking up currants from the floor, and she could reach out and catch a passing fly.[6]

But I continued to have a nagging doubt about what had been achieved: my hunch was that despite her manifest ability Helen remained to the end *unconscious* of her own vision. She never regained what we – you and I – would call the sensations of sight. Do not misunderstand me. I am not suggesting that Helen did not eventually discover that she could after all use her eyes to obtain information about the environment. She was a clever monkey and I have little doubt that, as her training progressed, it began to dawn on her that she was indeed picking up 'visual' information from somewhere – and that her eyes had something to do with it. But I do want to suggest that, even if she did come to realise that she could use her eyes to obtain visual information (information, say, about the position of a currant on the floor), she no longer knew how that information came to her: if there was a currant before her eyes she would find that she knew its position but, lacking visual sensation, she no longer *saw* it as being there.

It is difficult to imagine anything comparable in our own experience. But perhaps the sense we have of the position of parts of our own bodies is not dissimilar. We all accept as a fact that our brains are continuously informed of the topology of the surface of our bodies: when we want to scratch an ear we do not find ourselves scratching an eye; when we clap our hands together there is no danger that our two hands will miss each other. But, for my own part, it is not at all clear *how* this positional information comes to me. If, for example, I close my eyes and introspect on the feelings in my left thumb I cannot identify any sensation to which I can attribute my knowledge of the thumb's position – yet if I reach over with my other hand I shall be able to locate the thumb quite accurately. I 'just know', it seems, where my thumb is. And the same goes for other parts of my body. I am inclined therefore to say that at the level of conscious awareness 'position sense' is not a sense at all: what I know of the position of parts of my body is 'pure perceptual knowledge' – unsubstantiated by sensation.

Now in Helen's case I want to suggest that the information she obtained through her eyes was likewise 'pure knowledge' for which she was aware of no substantiating evidence in the form of visual sensations. Helen 'just knew' that there was a currant in such-and-such a position on the floor.

This, you may think, is a strange kind of hypothesis – and one which is in principle untestable. Were I to admit the hypothesis to be untestable I should be reneging on the whole argument of this chapter. The implication of such an admission would be that the presence or absence of consciousness has no consequences at the level of overt behaviour. And if consciousness does not affect behaviour it cannot, of course, have evolved through natural selection – either in the way I have been arguing or any other. What, then, shall I say? If you have followed me so far you will know my answer: I believe that Helen's lack of visual consciousness would have shown up in the way she herself conceived of the visually guided behaviour of other animals – in the way she did psychology. I shall come back to this in a moment; I think you will be more ready to listen to me if I first refer to some remarkable new evidence from human beings.

In the last few years Weiskrantz and his colleagues at the National Hospital, and other neurologists in different hospitals around the world, have been extending our findings with Helen to human patients.[7] They have studied cases of what is called 'cortical blindness', caused by extensive destruction of the visual cortex at the back of the brain (very much the same area as was surgically removed in Helen). Patients with this kind of brain damage have been described in most earlier medical literature as being completely blind in large areas of the visual field: the patients themselves will say that they are blind, and in clinical tests, where they are asked to report whether they can see a light in the affected area of the field, their blindness is apparently confirmed. But the clinical tests – and the patients' own opinion – have proved to be deceptive. It has been shown that, while the patients may not *think* that they can see, they are in fact quite capable of using visual information from the blind part of the field if only they can be persuaded to 'guess' what it is their eyes are looking at. Thus a patient studied by Weiskrantz, though denying that he could see anything at all in the left half of his visual field, could 'guess' the position of an object in this area with considerable accuracy, and could also 'guess' the object's shape. Weiskrantz, searching for a word to describe this strange phenomenon, has called it 'blindsight'.

'Blindsight' is what I think Helen had. It is vision without conscious awareness: the visual information comes to the subject in the form of pure knowledge unsubstantiated by visual sensation. The human patient, not surprisingly, believes that he is merely 'guessing'. What, after all, is a 'guess'? It is defined in Chambers' Dictionary as a 'judgement or opinion without sufficient evidence or grounds'. It takes consciousness to furnish our minds with the sensations which provide 'evidence or grounds' for what our senses tell us; just as it takes consciousness to give our minds the subjective feelings which provide 'evidence or grounds' for our eating behaviour, or our bad temper, or whatever else we do with the possibility of *insight* into its reasons.

So if Helen lacked such insight into her own vision, how might it have affected her ability to do psychology? I do not think that Helen's particular case is a straightforward one, since Helen was already grown up when she underwent the brain operation and she may well have retained ideas about vision from the time when she could see quite normally. I would rather discuss the hypothetical case of a monkey who has been operated on soon after birth and who therefore has never in its life been conscious of visual sensations. Such a monkey would, I believe, develop the basic capacity to use visual information in much the same way as does any monkey with an intact brain; it would become competent in using its eyes to judge depth, position, shape, to recognise objects, to find its way around. Indeed, if this monkey were to be observed *in social isolation from other monkeys*, it might not appear to be in any way defective. But ordinary monkeys do not live in social isolation. They interact continuously with other monkeys and their lives are largely ruled by the predictions they make of how these other monkeys will behave. Now, if a monkey is going to predict the behaviour of another, one of the least things it must realise is that the other monkey itself makes use of visual information – that the other monkey too can *see*. And here is the respect in which the monkey whose visual cortex was removed at birth would, I suspect, prove gravely defective. Being blind to the sensations of sight, it would be blind to the *idea* that another monkey can see.

Ordinary monkeys and ordinary people naturally interpret the visually guided behaviour of other animals in terms of their own conscious experience. The idea that other animals too have visual sensations provides them with a ready-made conceptual framework for understanding what it 'means' for another animal to use its eyes. But the monkey who has been operated on, lacking the conscious sensations, would lack the unifying concept: it would no longer be in the privileged position of an introspective psychologist.

In the days when we were working with Helen, Weiskrantz and I used to muse about how Helen would describe her state if she could speak. If only she could have communicated with us in sign language, what profound philosophical truths might she have been ready to impart? We had only one anxiety: that Helen, dear soul, having spent so long in the University of Cambridge, might have lost her philosophical innocence. If we had signalled to her: 'Tell us, Helen, about the nature of consciousness', she might have replied with the final words of Wittgenstein's *Tractatus*: 'Whereof one cannot speak, thereof one must be silent.' Silence has never formed a good basis for discussion.

Too often in this century philosophers have forbidden the rest of us to speak our minds about the functions and origins of consciousness. They have walled the subject off behind a Maginot Line. The defences sometimes look impressive. But biologists, advancing through the Low Countries, should not be afraid to march around them.

4

HAVING FEELINGS, AND SHOWING FEELINGS

In our dealings with animals – in the farm, the laboratory, the zoo, and even in the chase – most of us give some thought to the animals' feelings. We try to decrease the animals' experience of pain; and sometimes (though less dutifully) we try to increase their experience of pleasure. But however well-intentioned we may be, no one could claim that we really know what we are doing. For we have no direct access to the animals' consciousness; indeed we cannot be sure that animals consciously feel anything at all. Appearances notwithstanding, it is logically possible that animals are (as Descartes believed) merely unconscious automata.

Few of us, however, can withstand appearances, and few of us would wish to. In practice it is what animals *look* like which counts. In the absence of any more rational criterion we tend to assume that animals are in fact rather like ourselves, and that if they do have feelings they will *advertise* them. We trust that an animal which is consciously feeling distressed will express it by, say, grimacing or crying, or that an animal which is consciously feeling contented will express it by, say, smiling or purring. In other words we leave it to the animal itself to let us know what is going on. And for any animal which fails to advertise its feelings we have literally no sympathy: we show negligible concern for a fish which neither squeals nor screws up its face when it is hooked through the lip by a fisherman – though we should have the fisherman arrested if he did the same thing to a kitten.

Without some such rule of thumb for guessing whether feelings are or are not present, life for the farmer, zoo-keeper or veterinary surgeon would certainly be very difficult. People who have daily to make decisions about the care of animals, balancing economic or practical constraints against a moral concern for the animals' well-being, cannot afford the luxury of philosophical uncertainty. They have to act as they see best on the evidence in front of them. When an animal shows *signs* of pain, they must quickly do something about it – even though a philosopher might choose to doubt that the animal is actually conscious of anything at all; and when an animal shows no such signs, they can relax – even though another philosopher might suggest that at that very moment the animal is actually in agony.

But the fact that people need a rule of thumb, the fact that the rule which seems easiest and most natural to adopt is one based on appearances, and the added fact that the use of this rule is generally regarded as morally acceptable, do not of course mean that we are right to use it. The assumption that if animals have feelings they will advertise them might be totally invalid; so might the assumption that if animals advertise feelings then they have them.

In so far as both of these assumptions rest on the idea that there is a direct causal relationship between having feelings and expressing them, they are poorly based. Fortunately there is another way of looking at the matter, and that is to regard the relationship between feeling and expression not as one of cause but rather as one of contingent correlation. There are strong evolutionary grounds for believing in the reality of such a correlation, that is, for believing that the capacity for having feelings has in fact evolved hand in hand with the capacity for expressing them.

The gist of the evolutionary argument is this. The capacity for having feelings (that is, for being consciously aware of them) is an evolutionary adaptation to *social life*. The capacity for expressing feelings (that is, advertising them) is also an evolutionary adaptation to social life. Hence in any animal in which the former capacity has evolved, the latter capacity is also likely to have evolved, and vice versa.

The idea that the capacity for expressing feelings is connected to social living is perhaps sufficiently uncontroversial for me to take it for granted. Expression is a form of communication, and communication is essentially a social act. It would simply not make sense for an animal to express its feelings to the air: to cry, to blush, to purr or raise its hackles in a situation where there was no other animal present and ready to respond to it. The whole point of an animal's being expressive is to influence (presumably to its advantage) the behaviour of another member of its social group – by bringing help, staving off attack, etc. Thus any animal which *does* have the capacity to express its feelings must, we can assume, have evolved that capacity in a social context: and the greater its repertoire of expression, the more complex is likely to have been the society from which it comes – the greater the degree of individual interaction, mutual understanding, social manipulation, co-operation and so on.

The other idea, namely that the capacity for being conscious of feelings is also connected to social living, is of course more open to debate. But I have argued the case for it in some detail in other chapters, and here I will give no more than a bare summary. I assume that any animal which lives in a complex social group needs above all else the ability to do what I have called 'natural psychology', the ability to model the behaviour of other members of its group. My suggestion is that consciousness, as a biological phenomenon, has evolved precisely to serve that need. For the true value of consciousness is that it gives the natural psychologist direct access to those psychological concepts – the concept of feeling pain, feeling fear, feeling contentment, etc. – without which it would find the task of modelling the behaviour of another animal impossibly difficult. Unless, for example, an animal has itself been consciously aware of the feeling of pain, resulting from its own injury, it could hardly begin to understand the behaviour of another injured animal. But apart from this role in helping a social animal to do psychology, consciousness has, as I see it, no value whatever. So if consciousness exists at all outside the human species, it most probably exists in those animals that live in sufficiently complex social groups, but not in those that don't.

Bringing these separate lines of argument together, we reach this conclusion: animals which do have the capacity to advertise their feelings have probably evolved in a social context which is likely also to have resulted in the evolution of consciousness, and animals which do not have that capacity have not.

Thus the rule of thumb we all use in practice – and would no doubt go on using whatever philosophers or scientists have to say – is not a bad one after all. A kitten which cries is probably feeling pain; a herring which doesn't isn't. But for the record, let me summarise my own view of the matter. The kitten doesn't cry *because* it's feeling pain. It cries to influence the behaviour of another cat. But the fact that it can expect its crying to influence another cat means that it has evolved in the context of a complex social group; the fact that it comes from such a complex social group means that it needs to be a natural psychologist; the fact that it needs to be a natural psychologist means that it has probably evolved the capacity for consciousness. And *that* is why we may have been correct at the beginning to assume that its crying means that it is consciously aware of pain.

5

CONSCIOUSNESS: A JUST-SO STORY

Biologists who have thought, but not thought enough, about consciousness will be found toying with two contradictory ideas. First – the legacy of the positivist tradition in philosophy – that consciousness is an essentially private thing, which enriches the spirit but makes no material difference to the flesh, and whose existence either in man or other animals cannot in principle be confirmed by the objective tools of science. Second – the legacy of evolutionary biology – that consciousness is an adaptive trait, which has evolved by natural selection because it confers some (as yet unspecified) advantage on the individuals who possess it.

Put in this way, the contradiction is apparent. Biological advantage means an increased ability to stay alive and reproduce; it exists, if it exists at all, in the public domain. Anything which confers this kind of advantage – still more, anything whose evolution has specifically depended on it – cannot therefore remain wholly private. If consciousness *is* wholly private it cannot have evolved. Or if it has evolved, it must in Hamlet's words be but private north-north-west; when the wind is southerly it must be having public consequences. If the blind forces of natural selection have been able in the past to get a purchase on these consequences, so now should a far-seeing science be able to.

Yet scholars will, I suspect, continue to tolerate the contradiction, paying lip-service both to the privacy and to the evolutionary adaptiveness of consciousness, until they are offered a plausible account of just wherein the biological advantage lies. At present, so far from having a testable hypothesis which we could apply to species other than our own, we lack even the bones of a good story about consciousness in human beings. I offer one here: a Just-So Story.

But first some pointers to what, in the context of this story, I take 'consciousness' to mean. I rely on there already being between us the basis for a common understanding. I assume that you yourself are another conscious human being; that you have a personal conception of what consciousness is like; that you have experienced, waking and sleeping, both its presence and its absence; and that having noticed the contrasts you have already formed some notion of what consciousness is for. I assume moreover that although you may never have had occasion to pronounce on it, you will not find it difficult to recognise someone else's pronouncements (mine, below) as true of your own case.

Provided, that is, you are in fact a conscious human being, and not as it happens an unconscious robot or a philosopher from Mars. Provided, also, that you have not been too much influenced by Wittgenstein. When Wittgenstein (in a celebrated passage I have already mentioned in Chapter 1) alluded to consciousness as a 'beetle' in a box – 'No one can look into anyone else's box, and everyone says he knows what a beetle is only by looking at *his* beetle . . . it would be quite possible for everyone to have something different in his box . . . the box might even be empty'[1] – he chose the name of a thing which has no obvious use to us, and thereby implicitly ruled out the possibility that the things in our several boxes might bear a functional resemblance to each other. But suppose the thing in the box had been called, let's say, a 'pair of scissors'. One person's pair of scissors might indeed look rather different from another's: long scissors, short scissors, scissors made of brass or steel. But scissors to be scissors have to cut. There is really no danger that what we both agree to call a 'pair of scissors' could in my case be, say, a jelly baby while in your case it is merely empty air.

From all I know about myself, what strikes me – and seems to give some kind of cutting edge to consciousness – is this. The behaviour of human beings, myself included, is in every case under the control of an internal nervous mechanism. This mechanism is responsive to and engaged with the external environment but at the same time operates in many ways autonomously, collating information, hatching plans, and making decisions between one course of action and another. Being internal and autonomous it also, for the most part, operates away from other people's view. You cannot see directly into my mechanism, and I cannot see directly into yours. Yet, *in so far as I am conscious*, I can see as if with an inner eye into my own.

During most of my waking life I have been aware that my own behaviour is accompanied by certain conscious feelings – sensations, moods, desires, volitions and so on – which together form the structure and content of my conscious mind. So regular indeed is this accompaniment, so rarely does anything happen to me without its being either preceded or paralleled by the experience of a conscious feeling, that I have long ago come to regard my conscious mind as the very same thing as the internal mechanism which controls my bodily behaviour. If I ask myself *why* I am doing something, like as not my answer will be framed in conscious mental terms: I am doing it *because* I am aware of this or that going on inside me. 'Why am I looking in the larder? Because I'm feeling hungry . . . Why am I raising my right arm? Because I wish to . . . Why am I sniffing this rose? Because I like its smell . . .'

Thus consciousness (some would say 'self-consciousness', though what other kind of consciousness there is I do not know) provides me with an explanatory model, a way of making sense of my behaviour in terms which I could not devise by any other means. And to the extent that it is successful, this is presumably because the workings of my conscious mind do in reality correspond in some formal (if limited) way to the workings of my brain. 'Hunger' corresponds to a state of my brain; 'wishing' corresponds to a state of my brain; even the organising principle of consciousness, my concept of my 'self', corresponds to an organising principle of brain states. Not that physiologists have yet come up with an analysis of brain activity along these lines. But that, for the moment, is their problem, not mine. As a child of the evolutionary process, whose ancestors have been in this business for many million years, I am, in relation to my own behaviour, like the ancient astronomer in Figure 1 who has found a way of looking directly at the wheels and cogs which move the stars across the heavens: the stars are my behaviour, the cog-wheels are the mechanism which controls it, and the astronomer peering in on them is I my self.

So what?

So, once upon a time there were animals ancestral to man who were not conscious. That is not to say that these animals lacked brains. They were no doubt percipient, intelligent, complexly motivated creatures, whose internal control mechanisms were in many respects the equals of our own. But it is to say that they had no way of looking in upon the mechanism. They had clever brains, but blank minds. Their brains would receive and process information from their sense-organs without their minds being conscious of

Figure 1. An astronomer breaking through the sphere of mere appearance and, by the power of his imagination, catching a glimpse of mechanisms in the reality beyond.

any accompanying sensation, their brains would be moved by, say, hunger or fear without their minds being conscious of any accompanying emotion, their brains would undertake voluntary actions without their minds being conscious of any accompanying volition . . . And so these ancestral animals went about their lives, deeply ignorant of an inner explanation for their own behaviour.

To our way of thinking such ignorance has to be strange. We have experienced so often the connection between conscious feelings and behaviour, grown so used to the idea that our feelings are actually the causes of our actions, that it is hard to imagine that in the absence of feelings behaviour could carry on at all. It is true that in rare cases human beings may show a quite unexpected competence to do things without being conscious of their inner reasons: the case, for example, of 'blindsight' (see Chapter 3), where a patient with a cerebral lesion can point to a light without being conscious of any sensation accompanying his seeing (and without, as he says, knowing how he does it).[2] But the patient himself in such a case confesses himself baffled; and you and I will not pretend that that would not be our reaction too.

Such bafflement, however, was one among the many things our unconscious ancestors were spared. Having never in their lives known inner reasons for their actions, they would not have missed them when they were not there. And whether we can imagine it or not, we should assume that, for the life-style to which they were adapted, 'unconsciousness' was no great handicap. With these animals it was their behaviour itself, not their capacity to give an inner explanation of it, which mattered to their biological survival. As the occasion demanded they acted hungry, acted fearful, acted wishful and so on, and they were none the worse off for not having the feelings which might have told them why.

None the less, these animals were the ancestors of modern human beings. They were coming our way. Though their lives may once have been comparatively brutish and relatively short, as generations passed they began to live longer, their life histories grew more complicated, and their relationships with other members of their species became more dependent, more intimate, and at the same time more unsure. Sooner or later the capacity to explain themselves and to explain others – to take on the role of a natural 'psychologist', capable of understanding and predicting his own and others' behaviour within the social group – would become something they could no longer do without. At that stage would not their lack of consciousness have begun to tell against them?

Not necessarily. At least not at first, and not to the extent that all that's said above implies. For inner explanations are not the only kind of explanations of behaviour. Debarred as our unconscious ancestors may have been from looking in directly on the workings of their brains, they could still have observed behaviour from outside: they could have observed what went into the internal mechanism and what came out, and so have pieced together an external, objectively based explanatory model. 'Why am I [Humphrey] looking in the larder?' Not, maybe, 'Because I'm feeling hungry', but rather 'Because it's five hours since Humphrey last had anything to eat' or 'Because Humphrey has shown himself to be less fidgety after a snack.'

In short, while our ancestors lacked the capacity to explain themselves by 'introspection', there was nothing to stop them doing it by the methods of 'behaviourism'. 'The behaviorist', wrote one of its first modern champions, J. B. Watson, 'sweeps aside all medieval conceptions. He drops from his scientific vocabulary all subjective terms such as sensation, perception, image, desire, purpose, and even thinking and emotion.'[3] And who better placed to follow this recommendation than an unconscious creature for whom such conceptions could not have been further from his mind? In fact, it is we conscious human beings who have trouble being hard-headed behaviourists: it is *we* who, as B. F. Skinner has lamented, 'seem to have a kind of inside information about our behaviour. *We* have feelings about it. And what a diversion they have proved to be! . . . Feelings have proved to be one of the most fascinating attractions along the path of dalliance.'[4]

Why, then, when ignorance of the inner reasons for behaviour might have been bliss, did human beings ever become wise? Adam, the behavioural scientist, might with Newtonian detachment have simply sat back and watched the apple fall; but no, he ate it.

What tempted him was a leap in the complexity of social interaction, calling in its turn for a leap in the psychological understanding of oneself and others. Suddenly the old-time psychology which was good enough for our unconscious ancestors – which may still apparently be good enough for Watson and for Skinner – was no longer good enough for their descendants. Behaviourism could only take a natural psychologist so far. And human beings were destined to go further.

At what point the threshold was crossed we cannot tell. But there is evidence that by three or four million years ago, and possibly much earlier, our ancestors had already embarked on what was in effect a new experiment in social living. Leaving behind the relatively dull life of their ape-like forebears – leaving behind their thick skins, large teeth and heavy bones, leaving behind their habitation in the forest and their hand-to-mouth existence as vegetarian gypsies – they sought this new life as hunter–gatherers on the African savanna. They sought it with stone tools, they sought it with fire; they pursued it with forks and hope. But above all they sought it through the company of others of their kind.

For it was membership of a co-operative social group which made the life of hunting and gathering on the plains a viable alternative to what had gone before. Life from now on was to be founded on collaboration, centred on a home base and a place in the community. This community of familiar souls would provide the context in which individuals could reap the rewards of co-operative enterprise, where they could benefit from mutual exchange of materials and ideas, and where (against all subsequent advice) they could become borrowers and lenders and then borrowers again – borrowers of time, of care, of goods and services. But most important, the community would provide them as they grew up first with a nursery and then with a general purpose school where they could learn from others the practical techniques on which the life of the hunter–gatherer depended.

But the intense social engagement which this new life-style entailed spelt trouble. For human beings would not, overnight, abandon self-interest in favour of the common good. And while it's true that each individual stood to gain by preserving the social system as a whole, each continued also to have his own particular loyalties – to himself, to his kin and to his friends. A society based, as this was, on an unprecedented degree of interdependency, reciprocity and trust, was also a society which offered unprecedented opportunities for an individual to manoeuvre and out-manoeuvre others in the group.

Thus the scene was set for a long-running drama of personal and political intrigue. Men and women were to become actors in a human comedy, played out upon the flinty apron-stage which formed their common home. It was a comedy which would be tragedy for some. It was a play of ambitions, jealousies, loves, hates, spites and charities, where success meant success in the conduct of personal relationships. And when the curtain fell it was to those who, as natural psychologists, had shown the greatest insight into human nature that natural selection would give the biggest hand.

Imagine now two different kinds of player, with very different casts of mind. One the traditional unconscious behaviourist, who based his psychology entirely upon external observation; the other a new breed of introspectionist, who took the short cut of looking directly in upon the workings of his brain.

The behaviourist starts with a blank slate. In the manner familiar to those who have followed the progress of behaviourism as a modern science, he patiently collects evidence about what he sees happening to himself and other people, he correlates 'stimuli' and 'responses', he looks for 'contingencies of reinforcement', he tries to infer the existence of 'intervening variables' . . . and thus, without prejudice, he searches for a pattern in it all.

This programme for doing psychology is not, let it be said, a hopeless one. It must have sufficed for our unconscious ancestors for many million years. It probably still suffices for many if not all non-human social animals alive today. With a bit of luck it might have sufficed for those who began to live the life of social human beings – had they but world enough and time, had there been no one else around with the gift of doing the job much better.

But now there *was* someone else around, and world, time and luck were all at once in short supply. An introspectionist had entered on the scene: someone who starts with a slate on which the explanatory pattern is already half sketched in. From earliest childhood the introspectionist has had the opportunity to observe the causal structure of his own behaviour emerging in full inner view: he has sensed the connection between stimulus and response, he has felt the positive and negative effects of reinforcement, he has been directly apprised of the intervening variables, and he has daily experienced the unifying presence of his conscious self.

In the first instance, certainly, the introspectionist's explanatory model applies only to his own behaviour, not to others'. But once a pattern of connections has been forced on his attention in his own case, the idea of that pattern will dominate his perception in other cases where the connections are not openly on show. Once an outer effect has been seen, in his own case, to have an obvious inner cause, the idea of that cause will help him to make sense of situations where the effect alone can be observed. Cover the face in Figure 2, and try *not* to imagine the face in Figure 3.[5] Notice that a fire in your own private hearth causes smoke to issue from your chimney, and try *not* to imagine that the smoke coming from the house across the road implies the presence of a fire within those walls as well.

Thus the introspectionist's privileged picture of the inner reasons for his own behaviour is one which he will immediately and naturally project on other people. He can and will use his own experience to get inside other people's skins. And since the chances are that he himself is not in reality untypical of human beings in general – since the chances are that, just as from house to house there is generally no smoke without fire, so from person to person there is

Figure 2. The face in Figure 3.

generally no looking in the larder without hunger, no running away without fear, no rage without anger, etc. – this kind of imaginative projection gives him an explanatory scheme of remarkable generality and power.

Let us return then to the age-old human play. Scattered among the population of unconscious behaviourists, there arose in time these conscious prodigies. Soon enough an unconscious Watson

Figure 3. The hidden face.

would find himself up against a conscious Iago, an unconscious Skinner would find himself laying suit to a conscious Portia . . . Natural selection was there to supervise their exits and their entrances.

It was clear where the story for the human species had to end. But for the rest of the animal kingdom? As the bias of my story must have shown, I am not yet convinced that any other species has followed the same path to consciousness as man. But studies of the social systems of other species are not far advanced, and studies of how individual animals themselves do their psychology are only now beginning.[6] It may yet turn out that there are, in fact, non-human species whose social systems rival the complexity of man's; it may yet turn out that individuals of those species are, in fact, making use of explanatory systems which bear the hallmarks of a mind capable of looking in upon the inner workings of the brain. Stories have been wrong before. The cat, we know, does not walk by itself. But the rhino? Nothing suggests that the rhino gets inside another rhino's skin.

Meanwhile, for the obvious candidates – the social carnivores, the great apes – there will be biologists who in fairness want to leave the question undecided. Undecided, but not undecidable. In medieval England a jury could bring in one of four verdicts at a trial: Guilty, Not Guilty, *Ignoramus* (we do not know), *Ignorabimus* (we *shall* not know).

'Ignoramus' may be a proper verdict for biologists. But if consciousness has evolved we shall know it by its works. 'Ignorabimus' would be a counsel of philosophical despair.

THE PATH TO SELF-KNOWLEDGE

Homo sum; humani nil a me alienum puto.
(I am a man; I count nothing human foreign to me.)
Terence, 2nd century BC

6
JOINING THE CLUB

For a natural psychologist, the idea of fear will originate when he himself *feels* fear, the idea of love when he himself *feels* love . . . But these feelings will not occur to any person *unconditionally*. It may be true, as I've argued in the preceding essays, that reflexive consciousness is the source of the psychological concepts in terms of which ordinary people think about behaviour. But for each of us the history of what has happened in *our* consciousness must depend on our own experience in the outside world. The range of psychological concepts which we can know about through introspection will therefore be, at best, only as broad as our experience is wide. It follows that, in so far as this is the way that people do psychology, any one person's understanding of human behaviour must be more or less constrained by what he himself has been through.

This is an important conclusion. It may however strike you as profoundly unsurprising. Few people – and few academic psychologists, whatever their persuasion – would dispute that for someone to have lived through a particular experience should make him a better judge of how any human being is likely to react in comparable circumstances. The conclusion would surely be the same if the 'someone' were, instead of a 'natural psychologist' of the kind I've pictured, a thoroughgoing behaviourist – if his understanding of behaviour were based wholly on external observation without the aid of introspection. The grounds for drawing the conclusion would in the latter case be different: the natural psychologist benefits from having lived through an experience because – among other things – it gives him first-hand knowledge of what it feels like, whereas the behaviourist benefits simply from being able to observe how in those circumstances a typical human being (himself!) behaves. But, either way, personal experience may be expected to prove an invaluable aid to doing psychology.

Let us however examine the case of the behaviourist a bit more closely. We must suppose that, since he denies himself the privilege of introspection, all he gets out of personal experience is the chance to observe what he himself does, how things turn out for him: his information can amount to little more than he would get from watching his own body on a television screen. But, that being so, it makes no substantial difference that the body he is observing is his own body – it might just as well be someone else's. Gilbert Ryle put it plainly: 'The sorts of things that I can find out about myself are the same as the sorts of things I can find out about other people, and the methods of finding them out are much the same . . . John Doe's ways of finding out about John Doe are the same as John Doe's ways of finding out about Richard Roe.'[1]

The behaviourist, then, should be able to learn as much (and as little) about human behaviour by being the audience to someone else's performance as by being the actor himself. The natural psychologist, however, has no such choice: he has himself to be on stage if he is to have access to the privileged information which conscious reflection on his own performance gives him; though he too may learn something from observing others, he cannot thereby learn all there is to learn.

So we come to a question of fact. Is personal experience necessary – and not merely sufficient – to ordinary people's understanding of behaviour? Does *insight* enlarge the mind in a way that outward-looking observation never could do?

My argument was based before largely on an appeal to a priori plausibility. But now that the role of personal experience has been called in question, the two rival accounts of how people do psychology come into open conflict over an issue which ought to be empirically resolvable. It should surely be possible to determine whether in the real world people learn more about psychology from observing themselves than they do – or ever could – from observing others.

Let me at once disillusion you: so far as I know there are neither experimental nor clinical studies which bear directly on this issue. That is not to say that the question is unanswerable; indeed I think the answer is already widely known. But the evidence for it lies not in the scientific literature but in the fund of our everyday beliefs and practices. Common opinion, common knowledge and common practice provide here as persuasive an authority as any; and the answer they hand down is that personal experience *is* of peculiar importance to people's education as psychologists.

Since part of my ground for asserting this to be the answer is that all ordinary people already believe it to be true, I do not intend to make heavy weather of convincing you. I shall simply give some examples, illustrating the way in which differences in personal experience may affect one person's understanding of another's behaviour. But my main purpose in these chapters is to explore a consequent question. How, if personal experience is so important, do people make sure that they get it – in the right way, at the right time, and of the right kind?

Because most people do, in the end, contrive to extend their own experience to a remarkable (yet often unremarked) degree, examples of *differences* in personal experience are not all that easy to come by. More often than not we lack the relevant 'controls' – individuals for whom we can be sure that the particular experience is missing. But one area where this problem is minimal is that of sexual experience.

Although the experience of sexual intercourse is not of course an all-or-nothing thing, it is near enough all-or-nothing for the world at large to draw a clear distinction between sexual initiates and virgins. Moreover, here as in few other areas a person's lack of experience may be publicly vouched for – for example by their bodily immaturity or by their dress. The absence of breasts or of beard, the wearing of a nun's habit or a monk's, are fairly sure signs of sexual inexperience.

Now, the gap in these virgins' personal experience ought, so my argument tells us, to leave a deep gap in their understanding of sexual behaviour, a gap which could not in principle be filled by the most assiduous reading of a sex manual or the most dedicated scientific observation of the behaviour of loving couples in the park. Is that the case? Certainly common wisdom has it so. No one would go for advice on the psychology of sex to a child who has not yet reached puberty; and the advice of the celibate Roman Catholic clergy on marital and sexual matters, though it is sometimes sought, frequently proves foolish and insensitive. But since no one does seek the advice of children and since what the clergy have to say remains for the most part a secret of the confessional, we have as usual no hard data based upon experimental tests.

Let us then be the subjects of a test ourselves. Consider these verses upon sexual love, in Dryden's translation from Book IV of Lucretius:

> [Now] when the Youthful pair more clossely joyn,
> When hands in hands they lock, and thighs in thighs they twine
> Just in the raging foam of full desire,
> When both press on, both murmur, both expire,
> They gripe, they squeeze, their humid tongues they dart,
> As each wou'd force their way to t'others heart:
> In vain; they only cruze about the coast,
> For bodies cannot pierce, nor be in bodies lost:
> As sure they strive to be, when both engage,
> In that tumultuous momentary rage,
> So 'tangled in the Nets of Love they lie,
> Till Man dissolves in that excess of joy.
> Then, when the gather'd bag has burst its way,
> And ebbing tydes the slacken'd nerves betray,
> A pause ensues; and Nature nods a while,
> Till with recruited rage new Spirits boil;
> And then the same vain violence returns,
> With flames renew'd th'erected furnace burns.
> Agen they in each other wou'd be lost,
> But still by adamantine bars are crost;
> All ways they try, successeless all they prove,
> To cure the secret sore of lingring love.

W. B. Yeats called these lines 'the finest description of sexual intercourse ever written'.[2] A description – yes, a depiction even. At one level it is pure behaviourism: Dryden the detached observer coolly painting in words a picture of what lovers do. But if that was all there was to it the poem would hardly deserve to be so celebrated, and especially not by Yeats. Sexual initiates should not after all need a poet to tell them what lovers *look* like in bed, and virgins, though they may find the information intriguing, might learn still more from a visit to the modern cinema.

But Dryden does not simply tell us what the lovers looked like. On another level his poem attempts to convey something to which the behaviourist must necessarily be blind, something which could never be achieved with pictures, the sense of what the lovers felt – a frustrated longing to lose their selfhood in each other. For Yeats the true purpose of the poem was to illustrate the difficulty of two becoming a unity: 'The tragedy of sexual intercourse is the perpetual virginity of the soul.'

To this deeper level of meaning, who responds? What does someone who has never himself experienced loneliness in the arms of a sexual partner make of Dryden's dry commentary on the lovers' oceanic exploits? How is the uninitiated to comprehend 'the secret sore of lingring love'? The words, if they register at all, may seem to him mere cynicism; worse, cynicism inspired by envy of another's obvious pleasure. Cynicism in a way they are, but never mere. To the reader, *empowered by his or her personal experience to understand them*, Dryden's remarks are addressed as a salutary reminder and as an omen. On that 'secret sore' past relationships have foundered, and future ones will. The poem, like all good poetry, depends for its effect on lighting a taper in the reader's memory. But it cannot illuminate what is not there.

It may seem absurdly scientistic to call Dryden an introspective psychologist, more so still to say that when he refers, obliquely or directly, to the lovers' private feelings he is attempting to set up an explanatory model of their actions. But that, in the end, is what he is and what he does; and he does so through reference to concepts which, because of people's differing experience, cannot be universally intelligible. His poem is, for sure, a brilliant description of publicly observable behaviour, and as such is available to anyone; but it also presents a psychological hypothesis about the lovers' state of mind in a language which speaks only to those in whom the relevant concepts have been planted by their own experience. It takes a thief to catch a thief – and an intimate of his own consciousness to catch the intimations of consciousness in others.

I have stressed the exceptions, yet in reality sexual intercourse is an area of behaviour where few adult men and women are total strangers. Most will have understood that poem, and more still will understand the example I give next. Here, in a passage from the Song of Solomon, is a palliative to the Dryden, an evocation of a gentler, more generous phase of loving where the 'tragedy' – if it is to come – is as yet unrecognised. The voice is a man's; then his woman answers him.

> 'A garden inclosed is my sister, my spouse; a spring shut up, a fountain sealed. Thy plants are an orchard of pomegranates, with pleasant fruits' . . . 'Let my beloved come into his garden, and eat his pleasant fruits . . . I sleep, but my heart waketh: it is the voice of my beloved that knocketh, saying, Open to me, my sister, my love, my dove . . . My beloved put in his hand by the hole of the door, and my bowels were moved for him.'

If these verses stir memories of your own past feelings for a sexual partner, spare a thought for those less privileged who even now may be labouring with some immaculate misconception of the horticultural images. Spare a thought for the Church Fathers who in the Authorised Version of the Bible summarised this passage of the Song: 'Christ setteth forth the graces of the church; the church prayeth to be made fit for his presence.' Like that strangely-labelled door which confronted Steppenwolf in Hermann Hesse's novel, the imagery here is ultimately '*Not for everybody*'. The key, once more, lies with our own experience. While the majority of us may be fortunate enough to enter into the meaning of the verses, none the less each of us enters alone and with a key which he himself has cut.

So we understand – most of us – these poems, and more importantly we understand the behaviour of real lovers, because, to put it unromantically again, we have most of us acquired through personal experience the concepts which allow us to model human sexual behaviour. Yet if the commonality of sexual experience brings the natural psychology of sex within range of the majority of ordinary people, by the same token the rarity of certain kinds of experience must have the majority, in other areas, psychologically inept. Let me turn to an area which most of us do *not* understand, namely the behaviour of people who are ill.

Among the Ndembu of Zambia curative rituals for those who are troubled with sickness or misfortune are performed by 'doctors' at great public gatherings. These doctors' chief qualification for their role is that they themselves should previously have suffered and been cured: 'Doctors or diviners reply to the question "How did you learn your job?" by the words "I started by being sick myself".'[3] The expectation that doctor and patient shall have shared the same experience is dramatically expressed in certain less common rituals where 'the doctor gives medicine to himself as well as to the patient and both give way to paroxysms of quivering, very unpleasant to behold'.

In our own society we do not insist that our doctors shall have been sick themselves. True, we should regard with suspicion a doctor who had never himself had a headache, never vomited or had a fever. Still, we confidently allow our bones to be set, stomach ulcers to be excised or insomnia suppressed by doctors who have never suffered those particular disorders. We expect, it seems, no more *insight* from a doctor into sickness than we expect insight from a policeman into what it feels like to be burgled or from a fireman into what it feels like to have one's home go up in flames.

We expect no more and get no more. Western medicine is, in reputation and in fact, a medicine of the body rather than the person. As often as not the patient is discharged from the clinic with a remedy for his bodily symptoms but with little advice given to him, or to his family, on how to cope with the 'abnormal' behaviour which sickness so frequently entails. Someone who is sick is, after all, likely to behave in ways which in a healthy person would seem totally bizarre. A man with back pain, for example, will not only walk and stand most strangely, but he may desert his work, grumble all the time, and stop sleeping with his wife; likewise the sufferer from migraine, deafness, constipation or lung cancer will, each according to his illness, show some form of behavioural eccentricity.

Animals have been reported to react to the illness of another member of their species with outright aggression. People do not generally demonstrate so blatantly their lack of fellow-feeling for the sick, yet precious few show real psychological understanding. If my thesis is right it could hardly be otherwise: for few, whether doctors or laymen, are qualified by their own experience in this area to be natural psychologists. Most sorts of illness are, by definition, rare. They are not like sex; most of us have no personal experience of them. When we are confronted, say, by a person with migraine, those of us who have never ourselves known migraine are conceptual virgins, unable to imagine the peculiar fear, the auras, the pain and depression, the *feelings* which the patient himself considers to be the cause – and the explanation – of his own untypical behaviour.

In those circumstances where should – where does – the patient turn? To fellow-sufferers if he can find them. People with back pain, for example, will seek out and talk to others who have experienced back pain: witness Professor Steven Rose, who tells how, having been dismissed by his doctor with a prescription for painkillers and bed-rest, he went on to discover through encounters with a series of strangers the 'subjectivity of collective back pain wisdom and its resultant companionship'.[4] In a parallel search for sympathy (meaning not merely condolence but intersubjectivity) rheumatics talk to rheumatics, asthmatics to asthmatics, migranoids to migranoids, dyspeptics to dyspeptics and so on. Indeed societies are formed precisely to foster this kind of communion, especially where the illness is something shameful or inhibiting. Thus we have 'Alcoholics Anonymous', 'Gamblers Anonymous', 'Depressives Associated' – a freemasonry of the psychologically rejected.

Unfortunately, however, the counsel of fellow-sufferers cannot always be available even to those who seek it. Nowhere is the psychological isolation of the sick more obvious or more tragic than in the case of patients who are terminally ill. Someone who is dying in hospital of cancer has little chance of finding among his would-be counsellors anyone else who has shared the experience of imminent death. As a psychotherapist aware of this problem, Dr E. Kübler-Ross attempted to educate herself and others professionally concerned with the terminally ill by conducting extensive interviews with dying patients. The interviews were observed by student therapists and subsequently analysed in a group seminar. Dr Kübler-Ross's book, *On Death and Dying*, succeeds in giving a remarkable picture of the patients' world. Yet she herself was under no illusion about the limitation of a psychology based solely on external observation. I quote the following passage from the book for the lesson of the last few sentences:

> The most touching and instructive change in attitude, perhaps, was presented by one of our theology students who had attended the classes regularly. One afternoon he came to my office and asked for a meeting alone. He had gone through a week of utter agony and confrontation with the possibility of his own death. He had developed enlarged lymph glands and was asked to have a biopsy taken in order to evaluate the possibility of a malignancy. He attended the next seminar and shared with the group the stages of shock, dismay, and disbelief he had gone through – the days of anger, depression and hope, alternating with anxiety and fear . . . He was able to talk about it in a very real sense and made us aware of the difference between being an observer and being the patient himself. This man will never use empty words when he meets a terminally ill patient. His attitude has not changed because of the seminar but because he himself had to face the possibility of his own death.[5]

How did this student learn his job? He started by being sick himself.

The Ndembu doctor, Dr Kübler-Ross's student, anyone who can claim personal experience as his starting-point becomes *de facto* a member of a special 'club'. The word club is A. Alvarez's, used in a different but related context in his study of suicide, *The Savage God*.[6] Alvarez opens his book with an account of his friendship with the poet Sylvia Plath at a time when she was about to make a third and final attempt on her own life. Alvarez too had tried to kill himself a few years earlier. 'Because', he writes, 'I too was a member of the club', Sylvia Plath could talk about suicide in a way in which she would not have talked to an outsider. She recognised, maybe, that their shared experience had given them a shared vocabulary of feeling, and that for neither of them did Boris Pasternak's words hold true: 'We have no conception of the inner torture which precedes suicide.'

How did Pasternak himself conceive of this torture? As follows:

> People who are physically tortured on the rack keep losing consciousness, their suffering is so great that its unendurable intensity shortens the end. But a man who is thus at the mercy of the executioner is not annihilated when he faints from pain, for he is present at his own end, his past belongs to him, his memories are his . . . But a man who decides to commit suicide puts a full stop to his being, he turns his back on his past, he declares himself a bankrupt and his memories to be unreal . . . The continuity of his inner life is broken, his personality is at an end. And perhaps what finally makes him kill himself is not the firmness of his resolve but the unbearable quality of this anguish which belongs to no one, of this suffering in the absence of the sufferer, of this waiting which is empty because life has stopped and can no longer fill it.[7]

We may, it's true, have no conception of this inner torture. But perhaps, on the evidence of this passage, Pasternak was not a part of his own 'we'. Alvarez recognises Pasternak as another member of the club.

Sylvia Plath made suicide the theme of her novel *The Bell Jar*. A *novelist* is in the most literal sense a 'modeller' of human behaviour, someone whose skill as a psychologist is required not simply to comprehend but to invent the things that other people do. Yet the novelist's road to psychological understanding is in principle no different from anyone else's. If personal experience is of crucial importance to the layman in his dealings with people in real life, so it is to the novelist in his dealings with characters in fiction.

It is in fact common practice among novelists to base their creative writing on their own experience. And it is, I think, a common and proper prejudice among critics to respect them for that practice. Admittedly there are good writers of fiction who spurn autobiography, but convincing portrayals of human behaviour have come again and again from the pens of writers who knew their characters from the inside: Dickens *as* David Copperfield, Tolstoy *as* Levin, Dostoevsky *as* the Idiot.

Thus Dostoevsky's Idiot is made to share with his creator the feelings that accompany an epileptic fit. Compare for example the Idiot's experience in the moments just before a fit with Dostoevsky's own. This passage is from the novel:

> He [the Idiot] was thinking, incidentally, that there was a moment or two in his epileptic condition almost before the fit itself when suddenly amid the sadness, spiritual darkness and depression, his brain seemed to catch fire at brief moments, and with an extraordinary momentum his vital forces were strained to the utmost all at once. His sensation of being alive and his awareness increased tenfold at those moments which flashed by like lightning. His mind and heart were flooded by a dazzling light. All his agitation, all his doubts and worries, seemed composed in a twinkling, culminating in a great calm, full of serene and harmonious joy and hope.

And this is from a letter written by Dostoevsky to Nikolai Strakhov:

> For a few moments before the fit, I experience a feeling of happiness such as it is quite impossible to imagine in a normal state and which other people have no idea of. I feel entirely in harmony with myself and the whole world . . .[8]

With this kind of insight, no wonder Dostoevsky is able to construct such an uncannily life-like model of his hero. Uncanny because the psychological concepts, the tools with which he does it, are derived from feelings which 'other people have no idea of'.

The Idiot, like Dostoevsky, also knows the experience of impending death. But here Dostoevsky uses his own experience in a different and most revealing way. In 1849 he had faced execution by a firing-squad, only to be reprieved at the last moment. In the novel the Idiot does not himself go to execution, but he tells a story of another man to whom it has happened, detailing the horror and wonder of the last minutes before the reprieve. 'I was so struck by that man's story', the Idiot says, 'that I dreamed about it afterwards, I mean, of those five minutes . . .', and it seems clear that through this dream experience he is allowed to enter into the convict's suffering and tell the story almost as if it had happened to himself. So what is Dostoevsky up to here? He is, on the face of it, using his own experience of execution to tell the story of a man (the Idiot) who is telling on the basis of *his* experience the story of another man (the convict) who is himself about to die. But it goes, I think, further than that. For, on another level, Dostoevsky appears to be using his own experience of creative story-telling to model a creative story-teller. Dostoevsky knows from the inside what it is like to tell a story from the inside: the Idiot becomes a novelist himself.

*

In earlier chapters I argued that the ability to do psychology is a biologically adaptive trait in human beings: in the course of evolution the best psychologists have proved to be the best survivors. We can now add another premise: the best psychologists are likely to have been those with the widest range of personal experience. A striking conclusion follows. If psychology means survival and experience means psychology, then experience means survival. So the extension of inner experience should itself be a biologically adaptive trait in human beings.

I have come with new force to my earlier question. How, if personal experience is so important, do people make sure that they get it – in the right way, at the right time, and of the right kind?

We can reject two answers as nothing like adequate: one, that people do it deliberately, the other, that they leave it to chance.

That they occasionally do it deliberately, I do not deny. Most people know at least in theory the value of novel experiences in 'broadening' their minds. It would be surprising if they did not sometimes go out and seek experience with that end in view. True, I do not think that mind-broadening is the *usual* motive for, say, people's first experiments in making love (carnal knowledge, so called, has intrinsic attractions over and above the insight it may give into what the psalmist meant by an orchard of pomegranates). Nor, for that matter, is it a common motive for attempted suicide (few novelists – Hemingway excepted – have ever taken themselves that seriously). But still, there are times when people do seek new experiences with the acknowledged purpose of helping themselves 'make sense', through introspection, of the behaviour of others.

The clearest cases are those where someone deliberately courts an experience which is obviously unpleasant. My mother once discovered that my young sister had swallowed twenty plum-stones, whereupon she herself swallowed thirty – in order, she said, that she should be able to understand my sister's symptoms. My father, when he was politically active, deprived himself of food for a week, in order that he might have a better idea of what it feels like to be a starving peasant. A colleague at Cambridge, studying a tribe of Amazonian Indians, joined the Indians in drinking a strongly emetic and hallucinogenic drug in order that, having experienced the sickness and the visions, he should have inside knowledge of what the Indians were about. I could multiply examples, and so no doubt could you.

Such acts of calculated self-instruction have, however, a rather artificial ring to them. They are the reasoned acts of intellectuals, scarcely to be expected of ordinary people, 'the greatest part of whom', thought the Scottish philosopher Thomas Reid, 'hardly ever learn to reason'. Yet every natural psychologist, if he is to make good use of the possibilities of introspection, must acquire a broad base of inner experience to which he can refer. Reid, discussing the faculty of sight, concluded that – since most people are too stupid to acquire it rationally – 'God in his wisdom conveys it to us in a way which puts all upon a level.'[9] Such faith in a democratic deity may be out of place. But we may still assume that insight, like sight, is not restricted to the cognoscenti: it too, in practice, gets distributed to the common man.

I do not deny that matters can be left, in part, to chance. Given sufficient time and opportunity, people might well expect to pick up most of the requisite experience simply by waiting passively for things to come their way. Wordsworth believed so:

> The eye – it cannot choose but see;
> We cannot bid the ear be still;
> Our bodies feel, where'er they be,
> Against or with our will.
>
> Nor less I deem that there are Powers
> Which of themselves our minds impress;
> That we can feed this mind of ours
> In a wise passiveness.

It is true that sooner or later, without seeking it, most people will find that they have, say, fallen in love, or been beaten in a fight, or suffered betrayal by a friend; they may even, if they are lucky (or unlucky, depending on how you look at it), find that they have accidentally swallowed twenty plum-stones. But what if the experience comes later rather than sooner? The costs of naïvety are likely to be heavy in terms of psychological misunderstanding. Too heavy: wise passiveness is as risky a way of feeding the mind as it would be of feeding the body.

The issue is so serious that it would be strange if it had been neglected by natural selection in the course of evolution. Given that the answer lies neither with deliberation nor with chance, I believe that biological mechanisms have in fact evolved to see to it that the job gets done by one of nature's standard tricks: self-help by an unwitting subject. At least three natural means exist for ensuring that people, like it or not, rapidly receive the conceptual education required to turn them into competent psychologists. They are (i) play, (ii) manipulation by the family, (iii) dreaming.

In the next chapter I shall be discussing in detail these biologically-based mechanisms for extending personal experience. Then I shall turn, in Chapter 8, to certain *culturally-based* phenomena which in effect – and maybe by design – achieve a similar end.

But, lest the case I am about to make should seem to imply that there is nothing to prevent every Tom, Dick and Sigmund from becoming a psychic polymath, let me end the present chapter on a humbler note. The chances are that any natural psychologist, with both nature and culture as his patrons, will win access to a wide range of experience and with it a corresponding range of mental concepts. Thus most people can count on becoming personally acquainted with most of the states of mind they are likely to encounter in others. Yet to this comforting rule there is at least one stark exception. There are certain areas of experience – everyday experience – of which many people are bound to remain ineluctably ignorant for no other reason than that their *bodies* set limits to the possibilities of feeling.

You will accept that a man born without legs, however good his education, cannot but lack insight into the condition of those with legs. Yes, and so what? Leglessness is a rare state of affairs, and it would be perverse to cite it in refutation of a general rule. If half the people in the world were legless that would be a different matter. Then let me cite what is indeed a different matter: the fact that half the people in the world are born without penises or testicles and the other half without vaginas, breasts or wombs.

The inability of people of one gender to comprehend the feelings of people of the other is legendary – and daily confirmed in the difficulty men have in understanding the psychology of women, women the psychology of men. Freud himself failed to come to terms with female psychology and his male heirs have done little better. In a book on menstruation, Penelope Shuttle and Peter Redgrove remark that 'male psychoanalysts have no knowledge of the special events which belong to women',[10] and the same complaint might be made in relation to sexual orgasm, pregnancy, childbirth and lactation, not to mention the many areas of behaviour where these basic reproductive activities stand as influential metaphors. The truth is that the sexes are divided by an insight barrier which, being founded on differences in anatomy and physiology, is more or less unscalable. The notice now reads '*Not for anybody*': not for the psychoanalyst and not for the natural psychologist either.

Except, that is, for the strange case of Tiresias. His story, as told by Ovid in the *Metamorphoses*, says it all. Once, when out for a walk, Tiresias saw two serpents in the act of coupling. When both attacked him, he killed the female with his staff. Immediately he was turned into a woman and became a celebrated harlot. As such he lived for seven years. In the eighth year he saw two serpents again, but this time killed the male, and so was returned to his former manhood. Later he was asked to settle a quarrel between Jupiter and Juno: Jupiter had claimed that women derive much more pleasure from sexual intercourse than men, and Juno had denied this as nonsense. Tiresias, as a uniquely qualified natural psychologist, was summoned to settle the matter from his personal experience. He answered:

> If the parts of love-pleasure be counted as ten,
> Thrice three go to women, one only to men.

Juno was so indignant at being proved wrong – incensed beyond measure by Jupiter's triumphant grin – that she condemned Tiresias to eternal blindness. Jupiter compensated him by granting him inward sight.

It was an appropriate gift of Jupiter's: God in his wisdom does, it seems, sometimes confer special privileges. Yet even if we are not all put on a level with Tiresias, nature, as we shall see in the next chapter, has contrived that no human being shall be conceptually starved.

7

DREAMING AND BEING DREAMED

We play, we become the playthings of our families, and we dream.

I proposed in the last chapter that play, manipulation by the family, and dreaming may represent three natural mechanisms by which human beings – as apprentice psychologists – become exposed to experiences which might otherwise have passed them by. 'Hold on,' you may be saying: 'I am prepared to hear you on the first two, but I shall take some convincing on the third. Dreams are unreal, everything which happens in them is imaginary. Are you going to suggest that people really gain anything of value through having imaginary experiences which arouse *imaginary* feelings?' No, I am not. But the question is a fair one, and one I should answer at the beginning of this chapter if I am not to get tripped up over play as well as dreams, since play, too, frequently has elements of 'unreality'.

There is, I admit, a seductive verbal symmetry in the idea that while real events arouse real feelings, imaginary events must necessarily arouse imaginary ones. But the distinction between real and imaginary feelings makes no sense. All feelings, whatever context they occur in, are internal creations of the subject's mind. Although they may be – and usually are – evoked by real events in the real world, it is not those events as such which evoke them but rather the subject's perception of and belief in the events. For a particular feeling to occur it is *sufficient* that the subject should have the appropriate perceptions and beliefs – that he should *think* the relevant events to be occurring.

For a child to feel, say, fear, it will be enough that he should *think* he is being chased by a crocodile: his fear will be the same whether the crocodile is a genuine crocodile, a pantomime crocodile or a dream crocodile conjured up by his imagination. Or for a man to feel, say, jealousy it will be enough that he should *think* his wife is being unfaithful to him: Othello was not saved from jealousy by the mere fact that Iago's story was not true.

'That's as may be. But I've got another question for you. The point of your argument is to establish that your three ways of extending personal experience are ways of introducing people to *novel* feelings. You're surely not going to say that people can experience through fantasy – waking or sleeping – a feeling *which they've never had before*?'

Yes, I am. For what is the objection? It can hardly be that a person cannot in principle have a feeling unless he already knows what it is like, since that would lead to the absurd conclusion that no one could have a novel feeling even in real life. Clearly for every feeling there has to be a first time, a time when that feeling is aroused instinctively through the first encounter with the relevant events. But if, later on, the same feeling will be aroused even when those events are purely imaginary, why should they not have been imaginary upon the first occasion, why should the subject not react instinctively to his own fantasies?

Perhaps the reason for being puzzled is the thought that unless someone has first come across a particular feeling in real life he has no way of knowing what kind of situation would engender it – and so can scarcely be in a position to invent the appropriate situation in fantasy. For example, someone who has never tasted strawberries is most unlikely to invent a strawberry and taste it in a dream, and correspondingly someone who has never tasted jealousy is unlikely to invent his wife's infidelity and taste that in a dream. I agree about strawberries but not about jealousy. But here the case rests on biology rather than logic. In the course of evolution, insight into the taste of strawberries has never been of any great concern to human beings, but insight into the feeling of jealousy has been. Consequently, there has been no pressure from natural selection to invent a fantasy strawberry, but there may well have been pressure to invent fantasy grounds for jealousy. As the puppet-mistress pulling the strings of human imagination, nature is – as I shall hope to show – quite as clever a manipulator as Iago.

'Very well. Suppose you are right. If feelings occur as instinctive reactions to the relevant situations, and if, what's more, the subject is innately disposed to generate certain of those very situations for himself, then why shouldn't he simply have innate knowledge of the feelings without having to go to all that trouble? Why should a person need to inform himself, in such a roundabout way, of information which must already be latent in his brain?'

It cannot, I think, be disputed that such a roundabout route is generally all that is available. To take a simpler example than jealousy, people will instinctively feel toothache if and when their teeth are damaged; information about toothache must therefore be genetically coded in the brain. Yet the fact is that a person – the owner of the coded brain – does not know the feeling of toothache until the damage has occurred. In rather the same way, the colour of a person's eyes is genetically coded, and yet the person does not know their colour until, say, he has looked in a mirror. There is no great paradox here, nothing actually illogical in the fact that someone should not have access to the information passed on by 'his own' genes. When Osip Mandelstam wrote 'I am both the gardener and the flower', he was – with poetic effect – confusing two logically distinct senses of 'I'.

Yet isn't there perhaps something strange in the fact that a person should not be able to take advantage of genetic information even when it might prove biologically adaptive for him to do so: when, as with feelings (but not eye-colour), innate self-knowledge might improve his chances of survival? Nature, one might think, could have managed things better. Yes, possibly she might. But only, I suppose, if there had not been important constraints on the way in which consciousness evolved.

To recapitulate the suggestion made in earlier chapters, I assume that the evolution of consciousness came relatively late, after the basic structure of the brain had already been laid down. Somewhere along the evolutionary path that led to modern man natural selection, acting to promote the best psychologists, brought about a remarkable advance in cerebral organisation, transforming a being who merely behaved into one who at the same time informed his mind of the subjective reasons for his own behaviour. This involved, I suggest, the evolution of a new brain, a 'conscious brain', parallel to the older 'executive brain'.

Before the conscious brain evolved, however, the executive brain already had innate programmes lodged in it, programmes relating for example to 'being frightened' or 'being jealous': thus preconscious human ancestors were acting out the physiological states of fear or jealousy long before they were in a position to have insight into them. Now, with the evolution of the conscious brain, the states of mind which correspond to such physiological states of the executive brain became potentially accessible to consciousness. But since the conscious brain did not itself contain the age-old genetic programmes, these states of the executive brain could only be revealed as and when they actually occurred. In other words it remained necessary that the executive brain should first be prompted instinctively into some new state before the corresponding image – a novel feeling – could be reflected in the mirror of consciousness and thus become part of a person's knowledge of himself.

So, for reasons which a logician might consider wholly contingent but a biologist might think unavoidable, the way to self-knowledge became self-exposure to the relevant experience – by whatever route might prove available.

Play

The role of play in extending subjective experience probably needs little elaboration here. Although theoreticians may argue about what exactly counts as play and which are the most important of its many functions, no one will question that human playfulness is a biological trait which manifests itself in one form or another of wilful adventure – adventure for the mind as well as for the body. Were we to ask a young child (or an adult for that matter) why he is doing whatever he is doing in play, he would naturally reply that he is simply having fun. But in the course of having fun he is – naturally – educating himself: throwing himself into new kinds of interaction with the physical and social worlds and thereby discovering new ranges of experience.

Most treatises on play have, it is true, emphasised the acquisition of objective rather than subjective knowledge. In a recent compendium space enough is given to the role of play in giving the player information about the external world – what things are like, how things are done, the ways that people act – but there is little explicit reference to its role in introducing him to internal states of consciousness.[1] But whatever the bias of the scientific literature, the phenomena speak for themselves: through playful adventure children stumble upon whole new realms of feeling – new sensations, new emotions, new desires. They discover what it feels like to be a life-tossed human being.

Look at a child playing hide-and-seek, tag, or king-of-the-castle, games which are more or less standard the world over. Feelings of anxiety, of satisfaction, of disappointment, of competitiveness, even perhaps of compassion – these and many other rarer and often unnameable ideas are being planted and tended in the child's virgin mind. One day, when the games are for real, the man fathered by that child will use his subjective knowledge of such feelings as the basis for a psychological model of his friends and adversaries.

The line between fantasy and reality in play is not an easy one to draw. Several authors have noted that, in play, potentially serious actions are rendered safe by being 'uncoupled from their consequences':[2] the child who is caught in hide-and-seek is not really punished nor is his finder really rewarded, no babies are actually born to 'mummies and daddies', and no one is really killed in 'murder'. By this criterion, play is not play unless the players acknowledge that they are 'only pretending' to be serious. Yet, despite my claim that fantasy may be as effective as reality in eliciting subjective states of mind, I do not want to give the impression that play never provides 'real' cause for feeling. For, needless to say, even pretence play is staged in the real world, and while the player may escape some of the consequences of his actions he cannot escape all of them. A child who falls over when playing cops and robbers may not be a real cop and may have no real reason to regret the robber's getaway, but he may really cut his knee. In general, although games may have no serious goal (other than to reach that goal!) they invite encounters with events which must be taken seriously: along the line, children in play strain and tire their bodies, they break skins, break toys, break the rules of the game and sometimes break each other's hearts, they strike bargains and cheat on and get cheated by their playmates . . .

Against the rather cosy view of play as a light-hearted and innocent activity, it needs stressing that play is not altogether 'safe'. I do not know the statistics, but I dare say that hospital records would show that many if not the majority of serious accidents to children occur in the course of 'having fun' (just as the majority of students' broken legs occur in 'sport') – and I am quite sure that the majority of serious provocations to tears, to grins of triumph, blushes of shame, abject apologies, and acts of spite and altruism arise through the intrusion of reality into the lives of children who are ostensibly at play. The child proposes, the world (hand-in-hand with the flesh and the devil) disposes.

But what of the truly unworldly, imaginary side of play? While it may be the case that all play has one foot in fantasy and the other in reality, some kinds of play lean much more heavily on fantasy than others. Perhaps, in an unusually nicely balanced case, a girl will run to bring her doll in from the rain, thereby protecting her *imaginary* baby from *really* getting wet; but, to cite two opposite extremes, a child in a play-fight is living in fantasy only in so far as he does not really dislike his adversary and (probably) intends him no real hurt, whereas a child who is attempting to avoid the bears which lurk in the lines between the paving-stones is living in reality only in so far as he has to watch his step. Provided the child believes in his situation, real or imaginary, he may potentially gain experience by play of either kind. Yet, because what happens in reality is in the end bounded by material circumstance and law, whereas what happens in fancy is relatively free, it is in fact the latter which promises the most liberal introduction to the vocabulary of feelings.

As testimony to the astonishing creativity of a child's waking imagination, I can hardly do better than quote at length the memoir of Simone de Beauvoir:

> The games I was fondest of were those in which I assumed another character; and in these I had to have an accomplice . . . At that evening hour when the stillness, the dark weight, and the tedium of our middle-class domesticity began to invade the hall, I would unleash my phantasms . . . At times I was a religious confined in a cell, confounding my jailer by singing hymns and psalms. I converted the passivity to which my sex had condemned me into active defiance. But often I found myself revelling in the delights of misfortunes and humiliation. My piety disposed me towards masochism; prostrate before a blond young god, or, in the dark of the confessional with suave Abbé Martin, I would enjoy the most exquisite swoonings: the tears would pour down my cheeks and I would sink into the arms of angels. I would whip up these emotions to the point of paroxysm when, garbing myself in the bloodstained shift of Saint Blandine, I offered myself up to the lion's claws and to the eyes of the crowd. Or else, taking my cue from Griselda and Genevieve de Brabant, I was inspired to put myself inside the skin of a persecuted wife; my sister, always forced to be Bluebeard or some other tyrant, would cruelly banish me from his palace, and I would be lost in primeval forests until the day dawned when my innocence was established, shining forth like a good deed in a naughty world. Sometimes, changing my script, I would imagine that I was guilty of some mysterious crime, and I would cast myself down, thrillingly repentant, at the feet of a pure, terrible and handsome man. Vanquished by my remorse, my abjection and my love, my judge would lay a gentle hand upon my bended head, and I would feel myself swoon with emotion. Certain of my fantasies would not bear the light of day; and I had to indulge them in

secret. I was always extraordinarily moved by the fate of that captive king whom an oriental tyrant used as a mounting block; from time to time, trembling, half-naked, I would substitute myself for the royal slave and feel the tyrant's sharp spurs riding down my spine.[3]

Torture, religious passion, the delights of misfortune and humiliation, exquisite swoonings, banishments into the forest, thrilling repentance – unlikely experience for a young girl. If Mlle de Beauvoir had simply waited in life's queue to collect whatever experience the soup-kitchen of her middle-class existence might have had to offer, she would no doubt have waited for ever before a tyrant's sharp spurs came riding down her spine. As it was, she contrived to feed her mind on a banquet she herself prepared.

No one waits in life's queue. Although not every child may be as inventive as the young de Beauvoirs, all forage for and cook up experience for themselves.

I shall postpone till later in this chapter a discussion of the extent to which the experience thus amassed through play is particularly pertinent to the conceptual education of a natural psychologist, and with it a discussion of the hand of nature in determining the form of particular games and fantasies. At this stage let me simply say that while I do not of course believe that nature should be held responsible for, say, the precise rules of the game of tag or the details of mademoiselle's day-dreams, neither do I believe that she can be totally exonerated. Although, on the surface, different forms of play may seem merely frivolous, opportunistic or accidental, there are underlying biological mechanisms at work which see to it that through play a child fills his conceptual tool-bag with a selection of instruments peculiarly adapted to his future needs as a psychologist.

To return to my earlier image, play is one of nature's most effective ways of ensuring that natural psychologists grow fat. But it is not the only way. While children in play may be counted on to cater largely for themselves, there are other contexts in which they need – or certainly get – more forcible feeding.

Manipulation by the family

Wide as the scope of play may be, there are limits to the range of experience that it offers. Children play when and if it pleases them to do so, they indulge in it freely and withdraw from it freely – and they will not, if they can help it, choose freely to have experiences which are in no way pleasurable. Although I have mentioned the ever-present possibility of accident in play (and Simone de Beauvoir has testified to the possibility of masochistic – if romantic – fantasies), the fact is that any child left wholly to his own playful devices would be liable to get a highly distorted sample of experience, a good deal of jam and not much vinegar.

As an up-and-coming natural psychologist, such a child would be the loser. The concepts essential to modelling the behaviour of others in a social group will frequently be related to feelings which are *disagreeable* to whoever has them: depression, rage, pain, grief, helplessness, panic, the withdrawal of love, the desire to do harm . . .

Given that children, rather than courting such experiences, will actively avoid them, their education in this area raises special problems. The solution, I suggest, has been to administer the vinegar like nasty medicine in a spoon. And the spoon, as often as not, is held by a member of the family.

The theory of kin-selection shows that, biologically, it is in the interest of family members to increase the fitness of their relatives in whatever ways they can (provided it involves no great cost to themselves). Ethologists have long realised that this is the reason why, both in man and animals, parents and elder siblings so often take a hand in the education of younger members of the family, giving them lessons in how to do things and being active partners in their play. But there has been very little recognition of how relatives and friends might help young children by *abusing* them 'for their own good'.

By abuse I do not at all mean punishment. When a child is punished he is punished for some specific misdemeanour, and the purpose of the punishment is to make him change his ways. But abuse, in the sense I refer to, will have no such acknowledged purpose. Indeed it may often seem gratuitous, spiteful or unfair. Yet its effect will be to extend the child's range of psychological concepts by showing him how it feels to be a victim of misfortune or ill-use.

Let me illustrate the principle with a happening I witnessed not long ago on the train to Cambridge. A woman sat opposite me in the carriage with her four-year-old daughter. The little girl asked her mother an innocent question. The mother pretended not to notice her. The girl repeated her question, adding plaintively

'Mummy, please tell me.' 'I'm not your mummy,' the woman said; 'your mummy got off at the last station.' The girl began to look anxious. 'You *are* my mummy. I know you're my mummy.' 'No I'm not. I've never seen you before.' And so this strange game, if you can call it such, continued until eventually the bewildered little girl broke down in tears. A wicked, heartless mother? I thought so at the time. But maybe it was an unfair judgement. That little girl was in the truest sense being taught a lesson, the lesson of what it feels like to be mystified and scared. She perhaps learned more in those few unhappy minutes than I myself have learned from the hundred books I've read on train journeys.

Lest you take this to be an isolated instance, which could only happen with a slightly crazy mother or perhaps with a mother on her way to Cambridge ('For Cambridge people rarely smile, Being urban, squat, and packed with guile', wrote Rupert Brooke), let me cite further examples of such unkind 'teasing' from around the globe.

J. P. Loudon, reporting from Tristan da Cunha:

I watched the following performance, or variations of it, on a number of occasions over a period of some weeks . . . The father played a game with the child which consisted in his repeatedly hiding the child's favourite toy in different places round the room. For a time the game would proceed happily. When, however, as usually happened sooner or later, the child began to complain or show signs of exasperation because he was growing tired or could no longer find the toy or reach it down from its hiding place, the father spoke sternly to him and, seizing the toy, would make a pretence of throwing it onto the fire or out of the window. If this produced further reaction, which it generally did, the father threatened the child with punishment and, if he continued to whimper or started crying, the father smacked him.[4]

Barbara Ward, from Kau Sai in the province of Hong Kong:

A group of small children is playing ball, a bigger boy comes along and kicks the ball away from them. The small chap, Ah Kam, picks it up again and goes off with it, but before he has gone very far an adult comes along and knocks it out of his hands . . . A bit later Ah Kam starts playing at 'cooking' – with scraps of straw for firewood and broken up pots for rice and dishes – with another group of his age mates. This game goes on quite successfully for about fifteen minutes, until along comes another adult who, with a sweep of his hand, knocks the pretence rice and the pretence pots and pans flying . . . Ah Kam falls to the ground in an apparently uncontrollable rage. He yells for some time. Nobody takes any notice of him except perhaps to look in his direction once or twice.[5]

Philip and Iona Mayer, from the Transkei in Southern Africa:

In what is called *thelekisa*, women will catch hold of the hands of two little boys, two or three years old, and make them hit each other in the face, until the children get excited and angry and start lunging out on their own account, scratching and biting for good measure. The women look on with loud laughter.[6]

Admittedly, in each of these examples the grown-ups, we are told, would justify their persecution of the children as being 'character-forming' (for the child presumably, rather than the adult). No doubt, along with Dr Arnold of Rugby School, they would be partly right – such treatment could not but influence the development of character, though whether for better or worse is not entirely clear. But my own interpretation and emphasis are different: a child thus ill-used is surely learning how to characterise, from his own experience, the behaviour of *others* when in trouble.

Manipulation, abuse, teasing – call it what you will – of children by family and friends is, I think, much more widespread than scientists have noticed or perhaps cared to admit. But we should not be shy of recognising its potential function. Children as apprentice psychologists need to know about fright, so grown-ups – most probably their parents – frighten them; they need to know about jealousy, so parents do things to make them jealous; they need to know about pain, so parents hurt them; they need to know about guilt, so parents contrive to catch them doing wrong.

Once the search is afoot for such phenomena, examples are not hard to find. You will probably know of some tucked away in your own personal history. But rather than recounting further anecdotal instances, let me turn to one of the most famous, most universal and most curious of examples. I mean the so-called Oedipus Complex, as classically described by Freud.

The Oedipus Complex is complex in more senses than one, and by no means a pure case of the manipulation of a child's experience by his parents. Yet it merits discussion in this context on three counts: (i) the parents, wittingly or not, are in effect the agents of the child's distress; (ii) the experiences the child encounters – of love and hate within the primary family – are of central importance to his later social understanding; (iii) the course of the emotional conflict between parents and child was thought by Freud to be instinctive, and there is every reason to suppose that it may indeed have major biological determinants.

I shall for simplicity (and perhaps for plausibility also) confine my account to the case of a male child. The story goes something like this (emphasis being given to the *child's* emotional experience). The mother loves the son, and the son *loves* the mother. Her love is affectionate, fond and caring but non-sexual, his love is *demanding* and *sexual*. The mother will not and cannot allow herself to become her son's exclusive property: at times she shows indifference to him – he responds to indifference with *longing*; at times she deliberately rejects him – he responds to rejection with *disappointment*, leading to *despair*; at times she demonstrates her preference for the father – he responds to that with *jealousy* and *hatred*. The father loves the son, and the son *respects* the father. Because the son respects the father he responds to his own hostile feelings towards him with *guilt*. And, inverting his own resentment, he feels *resented by* the father. The father will not in fact allow his own position with the mother to be usurped by the son and a competition develops between the two very unequal males; knowingly or not the father taunts the son by his show of authority and power and he threatens symbolically (or in rare cases literally) to castrate the boy. The son *fears* such extreme retaliation. Eventually the son resolves his intolerable dilemma by identifying with the father, thus allowing him to experience vicariously his father's privileged position.

This account is of course an abstraction, pulling out some of the major themes from a great variety of primary-family interactions. But whatever extravagant claims may have been made for it, and whatever scepticism they have engendered, it is now widely accepted that some such pattern of emotional entanglement is common enough in early parent–child relationships. Forget for the moment about the role the Oedipus Complex may play in Freud's theory of the 'super-ego': my point, here, is that the child at the centre of the tangle is being given a spread of subjective experience such as he could hardly hope to get in any other way.

Yet is it fair to regard it as an example of active manipulation of the child's experience by the parents? The role played by the parents is, admittedly, to some degree a passive, even accidental one: it is not the mother's purpose that her son should have sexual designs on her, and what else can she do but disappoint him, what else can the father do but come between them? But let us not be deluded into thinking that mother and father are innocent bystanders in the affair. For the facts are: (i) the mother may 'lead her son on' by, for example, playing with him in an explicitly sexual or erotic way (including, not infrequently, deliberate masturbation); (ii) the mother, when she subsequently 'jilts' her child, may be not merely insensitive but deliberately cruel (as exemplified, perhaps, by the woman on the train); (iii) the father, when he 'puts the child down', may both flaunt his own virility and at the same time belittle the child's: if he does not actually go so far as to threaten to cut off his son's penis, he worsts him in ways which may be symbolically equivalent by, for example, hiding his favourite 'toy' or knocking away his 'ball[s]' (as shown perhaps in the cases from Tristan da Cunha and Hong Kong).

Dreaming

'I dreamt I was captured by cannibals. They all began to jump and yell. I was surprised to see myself in my own parlour. There was a fire and a kettle was over it full of boiling water. They threw me into it and once in a while the cook used to come over and stick a fork into me to see if I was cooked. Then he took me out and gave me to the chief, who was just going to bite me when I woke up.'

Thus an American boy reported that he spent his night.[7] Despite my suspicions of the way parents treat children I cannot think that any father would be likely to stick a fork into his son to see if he was cooked; nor does it seem probable that even the most enterprising child would invent such an episode in play. Wondrous strange things happen in dreams. And therefore, as strangers, the natural psychologist should give them welcome.

Dreaming represents, I think, the most audacious and ingenious of nature's tricks for educating her psychologists. In the freedom of sleep the dreamer can invent extraordinary stories about what is happening to his own person, and so, responding to these happenings as if to the real thing, he can discover new realms of inner experience. If I may speak from my own case, I have in my dreams placed myself in situations which have induced feelings of terror and grief, passion and pleasure, of a kind and intensity I have not known in real life. If I did now experience these feelings in real life I should recognise them as familiar; more important, if I were to come across someone else undergoing what I went through in the dream I should have a conceptual basis for modelling his behaviour. Dostoevsky's Idiot, you remember, *dreamed* that he was going to face a firing squad, and through that experience was able to enter into the condemned man's state of mind.

Listen, with sympathy, to the accounts of a few more children's dreams (all Londoners, aged from seven to twelve):[8]

I dreamt a dustman put me in a box and took me in a cart, and brought me back to the wrong bed.

Last night I dreamt that I was a brave Knight . . . A big monster came running after the most beautiful lady I ever saw. I drew my sword and hit the monster on the back. He roared so loud that the lady screamed. I said 'I will soon slay this monster', and I hit him again and then I cut off his head, and he fell dead on the ground. The lady said 'You are a hero.'

I dreamed I was in a swimming-bath full of water. My form master stood on the edge, in a red bathing-suit, and pushed me away with a long pole whenever I tried to climb out. I swam round and round in despair.

I dreamt I was married and had a little girl at school, also I had a little baby. My husband was a soldier in France. My girl had long curly hair. She wore a white frock with a blue sash and white shoes and stockings. But one night as I got into bed I heard the maroons go off. My children were fast asleep, so I got the eldest up first. I dressed them and put on their hats and coats and then dressed myself. Then I ran down the tube. A little while later I heard it was all clear, and when I got home I had just put my children to bed when I heard a rat-tat. I went and opened the door, and my husband walked in. He was home on leave. He was an officer.

I dreamt that I was going to be washed. After I was washed, I was rung out in the mangle. Then I was hung on the line. I was hanging on the line when it started to rain. My mother took me in and ironed me. The iron was hot.

How much of this experience is truly novel, without precedent in the child's waking life, it is of course difficult to say. The circumstantial evidence suggests that some of it can hardly be anything but novel: nine-year-old boys simply do not kill monsters, nor do eight-year-old girls have babies by officers in France. But because so much waking experience (including play, waking fantasies and abuse by one's family) and, for that matter, so much dreaming occurs very early in a child's life, it could never be a simple matter to establish the originality of dreams. If physiological measures are anything to go by we may estimate that between birth and the age of one a human baby has spent about 1,800 hours in dreaming, an average of about five hours a day.[9] By the time a child can talk and tell us what he knows the distinction between what he has learned through dreams and through waking experience has probably been blurred for ever. But there is one area of experience where, as I indicated in the previous chapter, it is easier to be sure. I refer to the experience of sexual intercourse.

It appears to be the case that a good many women (though rather fewer men) reach adulthood without ever having experienced an externally induced orgasm. Does it sometimes happen that such uninitiated women discover the feeling of orgasm by *dreaming* of the situation which evokes it? The answer, although some might dispute it, seems to be yes.

I offer as suggestive evidence the celebrated 'ecstasy' of St Teresa of Avila. St Teresa describes in her autobiography how at the supreme moment of her life she dreamed of an angel:

> In his hands I saw a long golden spear and at the end of the iron tip I seemed to see a point of fire. With this he seemed to pierce my heart repeatedly so that it penetrated to my entrails. When he drew it out I thought he was drawing them out with it and he left me completely afire with a great love of God. The pain was so great that I screamed aloud, but simultaneously felt such an infinite sweetness that I wished the pain would last eternally.[10]

The possibility that St Teresa is merely recalling in disguise some genuine sexual adventure of her earlier life can be discounted. If we were dealing with another saint, St Augustine for example, there might be reasonable cause for doubt. But not, I think, with St Teresa, founder of the austere order of Discalced (unshod) Carmelites, who was born into a pious Spanish family, entered early into a nunnery, and was noted all her life for her devoutness . . .

However, quite apart from the question of originality, should we perhaps query the *truth* of feelings experienced during dreams? Sometimes it is not in doubt: when the schoolboy swam round and round the swimming-bath *in despair*, or when St Teresa was left by the angel *afire with love*, the affective response is patently appropriate to the dreamer's picture of events. Yet whether people always experience the feelings appropriate to the manifest content of their dreams is not entirely clear. Freud commented: 'If I am afraid of robbers in a dream, the robbers, it is true, are imaginary – but the fear is real. And this is equally true if I feel glad in a dream. An affect experienced in a dream is in no way inferior to one of equal intensity experienced in waking life.'[11] But he also noticed that sometimes the feeling experienced does *not* apparently match the manifest content of the dream: 'I may be in a horrible, dangerous and disgusting situation without feeling any fear or repulsion; while another time, on the contrary, I may be terrified at something harmless and delighted at something childish.' Freud dealt with this enigma by supposing that, in the paradoxical cases, the feelings of the dreamer are those evoked by the latent content of the dream – the hidden thoughts behind it rather than the distorted version of those thoughts which reaches consciousness.

If it were the usual thing for the dreamer's subjective experience to be inappropriate to what he imagines to be happening to him, dreaming would provide an anarchic education for the psychologist, and my present argument would be hard to sustain. But even Freud did not believe it to be usual. In drawing attention to the paradoxical cases Freud had a particular axe to grind; what is more, he himself felt no need to extend his theory to the case of young children, in whose dreams he considered there to be a much closer correspondence between latent and manifest content (the dreams of children being for that reason 'quite uninteresting compared with the dreams of adults').[12]

That the dreamer's feelings are generally true to the content of the dream is confirmed not only by what the dreamer later recollects but also by signs of physiological arousal. C. W. Kimmins reports that a young girl who dreamed of the death of her brother wept tears during her sleep.[13] Studies of penis erection show a positive correlation between specifically sexual dream content and sexual arousal.[14] Moreover the prima-facie validity of dream experience is testified by occasions when the dream is treated as if it were a part of waking life – by the dreamer or indeed by others. A boy of six dreamed that someone had given him a threepenny bit, and on waking searched his bed for it.[15] Lamia, in Roman myth, on hearing from a young man that he had in a dream enjoyed from her 'that pleasure which she had denied unto his waking senses', sued him for a reward, 'conceiving that she had merited somewhat from his fantastical fruition'.[16] I have myself not infrequently carried over from dreams feelings of tenderness or anger towards other people, based on fantasy encounters with them (which they – were they to know of them – might well have reason to resent).

Against what I am saying, it may be objected that dreams are all too frequently forgotten – and that they cannot therefore play a major role in a child's conceptual education. Undeniably, memory for dreams is surprisingly labile, and much potentially valuable material must be lost. But even if only a fraction is retained, that fraction – especially in children – is impressive. Kimmins, having recorded over 5,000 dreams from London schoolchildren, either in narrative or written form, comments:

> It will be seen by the children's accounts of their dreams that they have an abnormal power of graphic description of events in which they are intensely interested, such as those supplied by the dream material. This power so far exceeds their ability in ordinary essay-writing on topics selected by the teacher, and is, moreover, so much in advance of their general standard of achievement, that it would appear as if some fresh mental element had come into play.[17]

Children, it seems, may find the lessons of the night at least as memorable as the lessons of the day.

All things considered, I invite you to accept that dreams – being memorable, valid, original and strange – may provide the growing child with a rich feast of psychological ideas. But one question will be worrying any attentive dietitian, a question unresolved throughout this chapter: To what extent is the experience gained in the several ways I've mentioned especially designed to meet a would-be psychologist's real needs? How much of this experience is genuinely nourishing and how much of it mere roughage?

To be specific, given that what matters to a natural psychologist is the possibility of insight into the behaviour of real people in his particular community, what ensures that (i) a dreamer dreams *relevant* dreams, (ii) a player plays *relevant* games, and (iii) members of the family put children through *relevant* trials?

We must not disallow that there may in fact be no ensuring of relevance, and that the whole thing could be left to chance. Serendipity, although not the most reliable of strategies, would all the same be better than no strategy at all. If a dreamer, for example, were to dream at random, simply turning the kaleidoscope of the previous day's events to see what he came up with, he would quite likely come up with something of potential value, and he would certainly be more likely to do so than if he were not to dream at all. Mr Micawber, always hoping that 'something would turn up', flourished in Australia *in the end*.

But suppose there really were to be some guiding principle directing children along the path of greater understanding, what line might we expect that it would take? Well, it is not hard to guess what kinds of personal experience it would be good for a young psychologist – in an ideal world – to get acquainted with. They fall into the following (overlapping) categories.

1. Experiences which he does not know of already, and especially those which he as a particular individual might otherwise never get to know.

2. Experiences which he will not get to know of in reality until he has grown older.

3. Experiences which he observes *other* people to be going through and which are characteristic of the life of the community.

4. Experiences which, whether he has had occasion to observe them or not, are characteristic of human beings in general.

Bearing in mind these criteria, we may then seek evidence of biological design in dreaming, play, and manipulation by the family. I am already committed to the view that it exists. I shall try to make the case in relation to dreaming, although much of what I have to say applies with equal force to the other two mechanisms.

1. Many dreams do, as I've already indicated, concern events of which the child has no previous experience. Indeed it seems to be surprisingly rare for a child to dream about everyday occurrences, for example school activities. Kimmins comments: 'Although the normal child spends nearly half the day in, or associated with, the school, it is remarkable that so few dreams have any direct reference to the school . . . In the infants' school only about 1 per cent are of this type.'[18] Moreover, not only does the child not dream of familiar events, but he sometimes dreams of events which to him could never in real life be anything but unfamiliar. Poor boys dream of being rich, rich boys of being poor, cripples of being able-bodied, healthy children of being crippled.

This 'compensatory' element in dreams, noted by many authors, was something which especially intrigued C. G. Jung. In Jung's view the function of compensation was to 'restore our psychological balance by producing dream material that re-establishes, in a subtle way, the total psychic equilibrium . . . The dream compensates for the deficiencies of the personality.'[19] However that may be (and in so far as I understand Jung's view I don't discount it), I would suggest an alternative function. The dream will compensate for the dreamer's ignorance of experience from which he as an individual is excluded: by dreaming of what he is not, the dreamer gains insight into what other people are.

2. Children commonly dream about experiences which in reality must lie well into the future for them. Most obviously they dream of being 'grown-up' – getting married, having babies, holding positions of power or responsibility, and facing death.

If such *prospective* dreams have a function, it may well be to give the dreamer insight not only into what life will be like for himself when he is older but also into what it is now like for others already more grown-up. (Compare the Freudian theory that a young boy, when he resolves the Oedipus Complex by identifying with his father, is enabled among other things to see himself (the child) from the father's (adult's) point of view.)

3. When a poor boy dreams of being rich or a little girl of being a mother, the dream clearly involves an element of *imitation*. Imitation is ubiquitous in dreams. The dreamer observes (or hears of) something happening to someone else, and in the dream contrives to make it happen to himself. Thus the Idiot, characteristically, first saw an execution and then 'dreamed about it afterwards', the girl who dreamed of her soldier husband coming home on leave had observed her mother welcoming her father home from the war, and the boy who dreamed of rescuing a beautiful lady from a monster had evidently been reading of St George.

I would regard imitation as the single most effective way of directing the dreamer's fantasy towards experience which will prove of use to him as a psychologist. Given that his future task will be specifically to model the behaviour of others in *his* social group, there could be no better preparation than to mimic in fantasy what he actually observes going on around him. If people around him fight, then let him dream of fighting; if they dance and sing, let him dream of dancing and singing; if they humiliate each other, let him dream of humiliating and being humiliated. In that way his dream-experience will be guaranteed relevance to the local conditions of his own community: he will learn, as it were, the *native* vocabulary of feelings.

4. Mireille Bertrand has drawn my attention to a singular exam-

ple of a monkey's dream.[20] The monkey, a six-year-old male Circopithecus which lived as a pet with a Parisian family, began unexpectedly to 'call' during its sleep. The calls were recognisably 'leader calls', loud barks of a kind which in the wild are made by young males when – at the age of about six – they leave their natal group. The pet monkey had lived with a human family since infancy and could never have heard another monkey calling in this way. The pet monkey's calls were given only when it was asleep and occurred in that deep stage of sleep which in people is associated with dreaming.

If this is what it seems, it suggests that this monkey may in fact have been rehearsing in its sleep the very experience which, though it could not have 'known' it, is sooner or later the lot of all male monkeys in the wild. The possibility that people too have instinctive dreams is one I have already alluded to. The evidence, admittedly, is far from solid. But when a person dreams of events which he cannot have observed and yet which represent episodes typical of the life-history of human beings, it is hard to avoid the conclusion that the dream has indeed been in some sense innately scripted. St Teresa's dream of intercourse with the spear-carrying angel provides one such possible example; and Jung, albeit with a different emphasis, has instanced many others.

For Jung they are examples of what he calls 'archetypal' dreams, dreams whose plots have been written by evolution into the 'collective unconscious' of mankind. In his words:

> Just as human activity is influenced to a high degree by instincts, quite apart from the rational motivations of the conscious mind, so our imagination, perception and thinking are likewise influenced by universally present formal elements ... The archetypes are the unconscious images of the instincts themselves ... They direct fantasy activity into its appointed paths and in this way produce in the fantasy-images of children's dreams astonishing parallels to universal myths.[21]

I cannot hope to summarise here the extensive research through which Jung claimed to have established the existence of particular archetypes. But, assuming the phenomenon to be genuine, I would propose a biological function for archetypal dreams: they will give the dreamer advance knowledge of certain universally significant human experiences, experiences of which he as a natural psychologist cannot afford to remain ignorant.

'We do not dream,' Jung said: 'We are dreamed.' We do not play, we are played. Our families do not manipulate, they are manipulated to manipulate us. All this is done by biological mechanisms designed to extend our insight into other people's feelings.

But human beings are greedy. Not content to let nature be their nurse, they have – as we shall see in the next chapter – devised a range of *cultural* mechanisms to stuff yet more experience into the growing psychologist's mind.

8

WHAT'S HE TO HECUBA?

Culture or no culture, people would still play and dream and wrangle with their families. But what is true for play is not, for example, true for 'plays'. A theatrical play, with its 'actors', its 'stage', its 'script', its 'dramatic conventions', could not exist outside the cultural framework which supports it. Like other cultural 'institutions', theatre originates with and is maintained by society rather than its individual members, it conforms to a pattern which is normatively sanctioned, and it depends for its survival on continuing tradition.

The activities described in the last chapter, universal as they are among human beings, are none of them in this sense institutions. But where nature leads culture often follows. And since nature herself has exploited most of the obvious possibilities for broadening experience, it is scarcely surprising that human cultures – in so far as they have been guided by the interests of their individual members – should have come up with similar solutions. Corresponding to the natural 'teaching aids' discussed in the last chapter, there exist institutional parallels the world over, supplementing, emulating and no doubt sometimes mimicking the natural activities. Natural play has its counterpart in institutional games and the provision of institutional playthings, family manipulation in institutional harshness and abuse, and dreaming in institutional fantasy.

The parallel, however, is by no means that straightforward. For, with the emergence of these culturally-based techniques which extend personal experience, there is a new and complicating factor to take into account. I mean the potential involvement of *spectators*.

Spectators are neither necessary nor desirable in the case, say, of a game of hide-and-seek, and there can never be a spectator of someone else's dream. In the case of institutional activities, by contrast, spectators may be not only invited but sometimes actually required. Twenty-two people play a game of football, a million see it on television; a dozen perhaps take part in a ritual circumcision, fifty stand around and watch; one writes a novel about ten characters, ten thousand read it . . . When spectators outnumber actors, the institution is presumably serving the interests of the former as well as the latter. But in what way?

I have up to this point insisted that if a person is to gain subjective experience he must, either in reality or fantasy, be himself at the centre of the action: observation of others is no substitute for personal participation. I am not about to go back on that assumption, but I do want to amend it. For there are in fact several ways by which non-participatory observation of others' behaviour can elicit – either immediately or later – new feelings in a 'mere' spectator.

For one thing, the spectator may, when it comes to it, be directly affected by the meaning for himself of what he observes happening to someone else. To take a couple of non-institutional and relatively trite examples, when a man finds his wife in bed with a rival his observation of *his wife's* behaviour may indeed elicit *his* jealousy, and when a lonely child sees his mother coming home the child's observation of *his mother's* behaviour may elicit *his* joy.

Here, however, we are talking of the observation of 'his wife' or 'his mother', and the spectator's subjective response clearly depends on his having an established relationship with the actor. If there is no such relationship, then, as a general rule, there's no response: I am not made jealous by the infidelity of someone else's wife, nor am I filled with joy by the sight of someone else's mother coming home. But an exception to this general rule is possible, indeed in certain cases likely: if circumstances conspire to give the spectator the *illusion* that he has a relationship to the actor – when in point of fact he has none whatsoever – his response to the actor's behaviour is likely to be every bit as strong as if the relationship were genuine.

Although this kind of illusion may occur spontaneously, certain cultural institutions positively foster it. They do so in three obvious ways. (i) The actors may be made to perform under the spectator's own 'banner', as if, like friends or kinsmen, they were doing him a personal service – such is the case, for example, with football teams, Olympic athletes and the British monarchy. (ii) The actors may be chosen from a special category of person with whom complete strangers will tend uncritically to assume the right or duty to a personal relationship – such is the case with infants, pretty women and (as we shall see) pet animals. (iii) The actors, enclosed within a special frame, may be subtly introduced to the spectator as intimates under conditions which encourage the 'willing suspension of disbelief' – such is the case with characters in drama and literary fiction.

I shall have more to say about all this. But let me now simply remind you where it leads. The spectator is duped into believing himself related to the actor, and hence will come to *mind* about the actor's doings and sufferings even though the actor's life has no bearing (in any other context) on his own. He will feel proud when his team's goalkeeper makes a good save and apprehensive when his pregnant Monarchess goes into labour, he will feel flattered by the attentions of a bunny-girl and sad at the death of a pet cat, distrustful of Lady Macbeth and captivated by James Bond. And when the forces that sustain the illusion of relationship combine, the effect on the spectator will be particularly powerful: thus the Christian's heart goes out to the *infant* Jesus, *our* Lord and Saviour, lying in the *story-book* manger, and a Londoner is greatly touched when the *elfin* Twiggy, with her *homely* cockney accent, blows him a goodnight kiss on *television*.

Yet, while in these ways the spectator may discover new feelings in himself, he will not be learning what it feels like actually to *be* the actor. He is reacting to rather than enacting the other person's situation, just as a husband reacts to rather than enacts his wife's infidelity. The jealous husband does not – it is probably the last thing he does – put himself in his wife's shoes. And, likewise, the reader of *David Copperfield* who is upset by Steerforth's elopement with Little Em'ly does not put himself in theirs. The reader feels protective towards Em'ly because he has become (under Dickens's guidance) a sort of elder brother to her, and he feels let down by Steerforth because he has become a trusty friend; but he has not become either of these characters themselves. At least that is the way I am describing it. But is that right? Do not spectators sometimes *identify* with actors and so truly learn what it feels like to be them?

Identification might mean one of several things. The best-known form depends on the spectator already knowing from his own past experience what it is like to be in the actor's situation. Hilaire Belloc's cautionary tale about Jim, 'who ran away from his Nurse and was eaten by a Lion', tests this kind of identification to – and beyond – its limit:

> Now, just imagine how it feels
> When first your toes and then your heels,
> And then by gradual degrees,
> Your shins and ankles, calves and knees,
> Are slowly eaten, bit by bit . . .

Most readers, when they try to imagine this, will fail when the lion gets past their toes. 'Sympathy' of the sort that Belloc is demanding is, as I discussed in Chapter 6, open only to those who are already members of the experiential club. As such, this kind of identification cannot in principle be a source of *new* insight into other people: the spectator may share his experience with the actor, but he does not learn it from him.

Most examples of so-called 'vicarious experience' are in this class. For instance, the spectator who sees a hypodermic needle being thrust into someone's outstretched arm, or reads of this in a novel (or in this very sentence), will inwardly wince if, but only if, he can identify with the patient on the basis of his past experience. The same applies to the spectator who is sexually aroused by watching another person masturbate, or who shivers to see someone standing barefoot in the snow.

There is, however, another more roundabout form of identification, which does not require that the spectator shall *already* have experience of just what it is he's observing. I mean identification by means of what an engineer would probably call 'simulation'. When an engineer wants to explore how a machine is likely to behave in untried circumstances he may set up a computer model into which he simply feeds the 'initial conditions' and then lets the 'simulator' run to see what happens. So a spectator may be tempted to explore the behaviour of an actor simply by adopting the actor's initial position and then waiting to see what comes about. Observing another person masturbate, he tries it for himself; or observing another person standing in the snow, he tries that too . . . The identification now is undertaken *speculatively*: the spectator need have no idea in advance what feelings his – or the actor's – act entails.

Identification by means of simulation is clearly something very different from the kind of sympathetic identification mentioned earlier: there the spectator must already know from his own past experience what feelings the actor is going through; here he still has to find out.

Since simulation *is* a way of finding out – of learning something new – it constitutes another potentially important means by which mere observation of an actor can put a spectator on the path to discovering more about the possibilities of human feeling. Were it not for his observations of the actor, the spectator would have no incentive to experiment with those particular initial conditions, and so might never know just where they lead.

Given, however, that simulation generally takes place *after* the spectator has witnessed the actor's performance, the question arises: Where does it take place, and when? The answer, already anticipated in the previous chapter, is that the two most important forums for it are without doubt dreams and play. When in my earlier discussion I drew attention to the fact that people commonly weave their fantasies around what they have seen happening to others in their social group, I was interested in this kind of imitation as a means of encouraging insight into the behaviour of the local populace. I did not at that point stress the role of cultural institutions in providing models for speculative fantasy (although Simone de Beauvoir and one or two of the quoted dreams provided sufficient testimony of it). Let me stress it now. Institutions – sports, rites, plays, stories – clearly can and do excite in their spectators the same tendency to imitation. Indeed, since institutional actors often prove more attractive than ordinary people as models – a child is more likely to dream of being a circus acrobat than of being a milkman – certain cultural institutions may as it were provide 'super-stimuli'. C. W. Kimmins has noted the special power of stories:

> The influence of fairy stories on the dreamer, especially the girl dreamer, is very marked. Greatest at eight years of age, it has considerable effect throughout the school period. The book read just before going to bed affects the dreamer and is often referred to by an explanatory note following the record of a strange dream, in which the child takes the part of one of the leading characters. Similar explanations are also given for what are evidently cinema dreams.[1]

I shall have more to say about this as well. But it is time to move on from a general discussion of spectator involvement to consider some specific institutions. The emphasis of the last few pages has been upon how spectators benefit from observing actors, but we must keep in mind that it may sometimes be the actors who have most to learn. I shall look at three kinds of institution: pet animals, initiation rites, and drama.

Institutional playthings: the case of pet animals

In the United States of America there are, according to recent estimates, 33.4 million pet dogs and 33.6 million pet cats, one dog and one cat for every six persons; in France it is one dog and one cat for every seven persons, in Britain one dog for every nine and one cat for every twelve. The pattern is similar across the whole of Western Europe. Half of all households have a pet animal, and two-thirds of households with young children do so.[2]

Although pet-keeping might appear to be entirely the responsibility of individuals, it has all the hallmarks of a cultural institution. Without society's recognition, facilitation and encouragement, pet-keeping as we know it would not survive. As an institution it has, however, been given astonishingly little attention by social scientists. No one seems to have noticed that in the United States there are nearly as many cats and dogs as there are televisions. The effects of television have been minutely researched and documented, but the effects of pets still remain virtually unanalysed.

If the analysis were to be done it would surely reveal that pets, far from being functionless adornments, provide a variety of services and comforts to the people living with them. One of the most important, I believe, especially where children are concerned, is to widen the personal experience of their owners.

I have headed this section 'Institutional playthings'. But 'plaything' is actually too weak a word for pets in the role I have in mind. I should have said 'player'. Or, since I do not mean simply a player of games but a player of dramatic parts, I should have said 'performer'. Admittedly pets are sometimes used as passive playthings, and they are often used as active playmates. But their most significant role, outside the context of ordinary play, lies in the way they perform as actors on the household stage. We should regard them, perhaps, as something like court jesters or Shakespearean fools, introduced into the bosom of the family to act out before a human audience a unique and uniquely affecting tragi-comedy. Although television may stand comparison with pets statistically, as a means of engaging people's hearts and minds it is hardly in the same league.

People observe – and react to – the show that pet animals put on, and thereby, as with other shows, they discover in themselves new possibilities of feeling. Pet animals, as actors, are in fact peculiarly well qualified for their role in thus teaching human beings about themselves. They are qualified, first, by the nature of what they do and what befalls them; and second, by the fact that the human spectator is so easily seduced into thinking that he has a personal relationship with them.

Let me begin with the nature of the relationship. I shall narrow the discussion to the case of dogs and cats as pets, although rabbits, hamsters, even budgerigars are in several ways similar.

It is a commonplace that people treat dogs and cats as if the animals were themselves human beings, and were thus fit to be partners in a human relationship (as friends, mates, children or whatever). The fact that the animals are able to sustain this illusion is of course no accident; indeed their very existence may depend on it, for they have often been selected and bred with that quality in view.

The resemblance of dogs and cats to human beings works at several levels. To start with, they look somewhat like them. They have expressive, mobile faces, expressive voices and expressive gaits. Their expressions are easy for people to read and are readily translated into human terms. They show what appear to be human emotions (happiness, contentment, sadness, jealousy . . .), and they seem to understand such emotions in people. They enjoy human pleasures (comfortable chairs, warm fires . . .). They eat human foods (meat, milk, fish . . .), and, like humans, they prefer it cooked. They suffer from human ailments (sickness, diarrhoea . . .). They show affection in human ways (by licking, nuzzling, caressing . . .), and they welcome the return of affection in like manner. They are highly sociable (especially towards human beings themselves), and yet, like humans, they also show considerable discrimination.

Besides resembling humans in these general ways, they often specifically resemble human infants. Konrad Lorenz listed the 'key features' which elicit from people innate patterns of child-care and affection.[3] Among them are: a small body and a large head, a flattened nose, large eyes, a prominent forehead, short thick limbs, a rounded body, a soft cuddly body surface. Cats, with little selective breeding, do well by these criteria; and several kinds of dogs, notably the Pekinese, have been artificially bred to meet them – their noses squashed, fur lengthened, limbs shortened, bodies rounded and reduced in size. Moreover, in addition to their infantile looks, the animals have been selected for their infantile behaviour. Dogs have been 'neotenised', so that they retain puppy-like characteristics – playfulness, subordinance, dependency – well into adulthood.

A human spectator who, taken by these attributes, believes that he has a personal relationship with the pet animal is, however, in large measure being deluded. I do not mean that there can be *no* genuine relationship; but, whatever it is, it is not a relationship between *persons*. The dog or cat, no matter how good a mimic, is ultimately not another human being, and the implicit belief that it *is* a human being can only lend to the relationship an illusory biological or social significance. Someone who believes, for example, that there is mutual love between him and his cat or mutual trust between him and his dog cannot – or rather should not – be using the words 'love' or 'trust' in the way that they may be used to describe the relationship between two people.

Yet, at an emotional level, the confidence trick unquestionably works. The feelings which pet animals engender in people are the very same feelings which might, in appropriate circumstances, be engendered by another person. And that gives pets, as household actors, a remarkable power to broaden the spectator's experience. For the fact is that pets – precisely because they are not themselves human beings – are not constrained by the rules which govern what happens to other members of the family. Pets do things and things happen to them which people, especially children, have never had the opportunity to witness being done by or happening to any other person about whom they care.

Pet animals are, for example, stupider, naughtier and dirtier than the majority of human beings – and a child, whose pet they are, may thus discover what it feels like to have superior knowledge, to make allowances for misdemeanours, to put up with and clear up after a less responsible member of the family, and to mete out corrective punishment. Pets are more dependent, more affectionate, more submissive than the majority of human beings – and the child may thus discover what it feels like to be needed, to be dominant, and to receive undeserved gratitude and admiration. Most important, pet animals live their lives *faster* than human beings. The round of birth, copulation and death comes to dogs and cats about seven times more frequently than it comes to humans. Thus a child's first involvement with these emotionally evocative events is likely to come through observation of a pet rather than of a human member of the family: the first copulation he sees may be performed by his dog on the garden lawn, the first birth the birth of puppies on the sitting-room floor, and the first death the death of his dog under the wheels of a car.

B. M. Levinson has written about how an animal's death affects a child:[4]

Today with the almost universal adoption of the nuclear family and practically the extinction of the extended family, deaths of relatives occur rarely and the child's first-hand acquaintance with death usually comes when a pet dies. In such a family a pet can be and often is a companion, friend, servant, admirer, confidante, toy, team mate, scapegoat, mirror, trustee and defender for the child. Death is an object of great concern to the child and he finds it difficult to accept the death even of a pet . . . The death of a pet is thought of as a punishment. A host of questions which philosophers have not been able to resolve then begin to trouble the child. Why was he punished? Will the pet come back? If not where does the pet's soul go? Uncomfortable thoughts about his parents dying and leaving him alone, uncared for and unloved, begin to trouble the child. The child begins to feel that maybe he has done something to cause the pet's death. Did he not sometimes wish that his pet, that big pest, would die? To a child, as we know, a wish is equivalent to an act. The child develops feelings of guilt and to assuage and abreact his guilt he engages in endless obsessive burials and disinterments of the pet's body. After the pet is repeatedly disinterred and the pet *is* dead, the child is reassured that the pet will not come back and bother him in his nightmares. Learning how to handle the painful mourning experiences of the death of his pet may strengthen the child's ego against greater losses in the future. If the family accepts the pet's death as a common loss or a common critical situation and the family mourns together for the loss, it will lessen the child's grief, helplessness, anger, hostility and fear, and teach him that sharing the feelings of loss with someone else will tend to decrease the feelings of guilt.

Levinson's psychoanalytic bias is perhaps unwarranted. But the feelings he describes are real, and what is significant is that they are the feelings people have about the death of *persons*. The child who is moved to grief, guilt and fear by the death of a pet dog is gaining insight into how people feel when a human loved one dies.

Although I, like Levinson, have talked of children, emotional concern with the fate of animals is not restricted to the young and gullible. Old ladies weep at the pets' cemetery in Hollywood, and kings have given horses stately funerals. Two thousand years ago Lesbia mourned for the death of her pet bird, and Catullus mourned for Lesbia:

> Lugete, o Veneres Cupidinesque,
> et quantum est hominum venustiorum.
> passer mortuus est meae puellae . . .

G. S. Davies' translation, in the style of Burns, recalls the tragedy:

> Weep, weep, ye Loves and Cupids all,
> And ilka Man o' decent feelin':
> My lassie's lost her wee, wee bird,
> And that's a loss, ye'll ken, past healin'.
>
> The lassie lo'ed him like her een:
> The darling wee thing lo'ed the ither,
> And knew and nestled to her breast,
> As ony bairnie to her mither.
>
> Her bosom was his dear, dear haunt –
> So dear, he cared na lang to leave it;
> He'd nae but gang his ain sma' jaunt,
> And flutter piping back bereavit.
>
> The wee thing's gane the shadowy road
> That's never travelled back by ony:
> Out on ye, Shades! ye're greedy aye
> To grab at aught that's brave and bonny.
>
> Puir, foolish, fondling, bonnie bird,
> Ye little ken what wark ye're leavin':
> Ye've gar'd my lassie's een grow red,
> Those bonnie een grow red wi' grievin'.

And, if that is bad enough, imagine the peculiar horror when one pet kills another. In *The Nun's Lament for Philip Sparrow*, written about 1500, John Skelton laments the death of a bird killed by a cat, in an elegy of more than thirteen hundred lines.

> When I remember'd again
> How my Philip was slain,
> I wept and I wailed,
> The tears down hailed;
> But nothing it avail'd
> To call Philip again
> Whom Gib our cat hath slain.
>
> . . .

> It had a velvet cap,
> And would sit on my lap,
> And seek after small worms,
> And sometimes white bread crumbs;
> And many times and oft
> Within my breast soft
> It would lie and rest.

Alas, poor Philip.

Alas, poor Yorick. The king's jester, too, must have had a velvet cap. 'I knew him, Horatio; a fellow of infinite jest, of most excellent fancy; he hath borne me on his back a thousand times . . . Here hung those lips that I have kissed I know not how oft. Where be your gibes now? your gambols? your songs?'

It is, of course, the greatest jest of all to die.

Institutional abuse: initiation rites

Plate XIII of Gregory Bateson's book *Naven* shows a photograph of a man wiping his bottom on another man's head.[5] The caption reads 'Initiation in Komindimbit [a village in New Guinea]: an initiator expressing scorn for a novice . . .'. Bateson's labrador dog may be seen at the right of the picture, apparently indifferent to the proceedings.

Whether the dog's lack of interest indicates that people have less to teach pets than pets have to teach people is not a matter that need concern us here: in discussing initiation rites my emphasis will be on the actor rather than the spectator. Whoever else may or may not have 'benefited' from this strange ceremony, we may be fairly sure that the novice, for one, was gaining novel personal experience.

I have already discussed the possible didactic function of 'spontaneous' ill-treatment of youngsters by grown-ups, and my case for institutional ill-treatment will be much the same: it gives the child privileged knowledge of what it feels like to be in trouble, to suffer hardship, pain, rejection, fear of mystery and fear of men.

Let me start with some examples. First, more from Bateson's *Naven*, in which he describes the initiation of youths into manhood as practised by the Iatmul of New Guinea. The captions to his photographs provide as good a summary as any:

> Plate V. The novice is separated from his mother.
>
> Plate IX. The initiators are waiting in two lines armed with sticks. The

novice will enter through the leaf screen at the back of the picture and will run the gauntlet of the initiators.

Plate X. The novice is lying prone on an inverted canoe clasping his mother's brother who acts as a comforter and 'mother'. An initiator of the opposite moiety to the novice is cutting the latter's back with a small bamboo blade. In the foreground is a bowl of water with swabs of fibre to wipe away the blood.

Plate XII. Bullying the novice. For about a week after his back has been cut, the novice is subjected every morning to a series of bullying ceremonials. He is made to squat like a woman while masked initiators maltreat him in various ways. The particular incident shown here is a recent innovation. The mask worn by the initiator, referred to as *tumbuan*, divines for theft. The cow's bone which is on the ground in front of the figure's knees is held up on a string in front of the novice's face. The *tumbuan* then says, addressing the novice in the feminine second person singular: 'Are you a child that steals calico?' He then moves his arm causing the bone to swing, which indicates an affirmative answer to the divination; and he slaps the novice. The same question is asked about yams, bananas, tobacco, etc., etc. In each case an affirmative answer is obtained and the novice is slapped.

Plate XIII A. An initiator expressing scorn for a novice by rubbing his buttocks on the latter's head.

Plate XIII B. A novice eating a meal. He is made to wash before he eats, and even then must not touch his food with his hands, but must pick it up awkwardly and miserably with a pair of bamboo tongs or a folded leaf. This was the youngest of the novices initiated in this ceremony [he looks about five years old]. The parts of his body which have been cut are wet with oil and the remainder have been smeared with clay.

Bateson comments in his main text:

The spirit in which the ceremonies are carried out is neither that of asceticism nor that of carefulness; it is the spirit of irresponsible bullying and swagger. In the process of scarification nobody cares how the little boys bear their pain. If they scream, some of the initiators go and hammer on the gongs to drown the sound. The father of the little boy will perhaps stand by and watch the process, occasionally saying in a conventional way 'That's enough! that's enough!' but no attention is paid to him . . . In the ritual washing, the partly healed backs of the novices are scrubbed, and they are splashed and splashed with icy water till they are whimpering with cold and misery. The emphasis is on making them miserable rather than clean.

The Iatmul of New Guinea are, it's true, a notoriously barbarous people, and not all societies which practise similar rites encourage or permit such outright sadism. None the less, even among gentler people, such as the Ndembu of Zambia, the novice is not spared anxiety and pain. In the Ndembu rite of circumcision, the circumciser, 'looking thoroughly menacing with red clay daubed on brow and temples and a red lourie feather in his hair', sings:

> I am the lion who eats on the path . . .
> Novice's mother, bring me your child
> that I may mistreat him.[6]

And if, compared to the Iatmul, the mistreatment which follows is mild, it must still be horrifying enough for a ten-year-old boy. 'Now drums began to thunder loudly to drown the cries of the novices. Kambanji was the first novice to be carried in. He looked really alarmed . . . and yelled out as Nyachiu seized him from behind and held him steady for the operation.'

A. van Gennep, early in this century, described in his classic work *The Rites of Passage* numerous forms of initiation as practised by peoples around the world.[7]

Among the Australian aborigines:

> The novice is secluded in the bush, in a special place, in a special hut, etc. . . . Sometimes the novice's link with his mother endures for some time, but a moment always comes when, apparently by a violent action, he is finally separated from his mother, who often weeps for him . . . In some tribes the novice is considered dead, and he remains dead for the period of his novitiate. It lasts for a fairly long time and consists of a physical and mental weakening . . . The final act is a religious ceremony and, above all, a special mutilation which varies according to the tribe (a tooth is removed, the penis is incised, etc.) . . .

Among the Ojibway Indians of North America:

> A sacred hut is built; the child is attached to a board and during the entire ceremony behaves as though he had lost all personality; the participants are dressed, painted, etc.; there is a general procession to the interior of the hut; the chiefs-magicians-priests kill all the participants and resurrect them one after the other . . .

Among tribes of the lower Congo:

> The novice is separated from his previous environment, in relation to which he is dead . . . He is taken into the forest, where he is subjected to seclusion, lustration, flagellation, and intoxication with palm wine, resulting

in anaesthesia. Then come the transition rites, including bodily mutilations and painting of the body; since the novices are considered dead during their trial period, they go about naked and may neither leave their retreat nor show themselves to men . . . They speak a special language and eat special food . . .

As in my discussion of family abuse in the last chapter, I have given these examples to drive home the point that institutional abuse is not an isolated or rare phenomenon. Indeed, given the spread of such rites among present-day primitive societies, we may guess that until the recent past they were the common lot of most of the world's children. Even contemporary American children do not by any means escape scot-free: 87 per cent of boys born in the USA were circumcised in 1976, and girls too were frequently circumcised at the beginning of this century.[8]

Parallels to brutal initiation rites survive in other 'modern' contexts, notably the army and the school. R. Lambert, in his book on English boarding schools, quotes three teenage boys:

'Well, they got me in the Prefects' room. They made me put my hand out, fingers spread on an old desk, then a prefect got a compass and began to stab the gaps between my fingers with the compass point, back and forwards, faster and faster. Then when he was doing it fastest, he shut his eyes. I was terrified. Thank Christ he didn't miss. They sometimes do and boys go to Matron – they daren't split – "My finger got hit by a nail" . . .'

'A bunch of them [the seniors] got us [the new boys] in the changing room and made us run "the gauntlet" – run under a tunnel of arms while they whacked or pinged us with their belts. The bastards, did it sting! If you yelped or cried, they called you yellow and wet and they didn't let you forget it quick either.'

'Every new boy who comes into the house has to go on the table. What is the point of putting a boy on a table and holding his arms and legs while the thugs [certain prefects] beat him up?'[9]

Good question. But, then, what is the point of tying a child to a board and pretending to kill him? Or of cutting a child's back with a bamboo blade and then bullying him for five days?

The answer is that, if there is any point, there are probably several. Anthropologists, noting that these rites are rites of *transition* whereby the novice changes his status in the eyes of society, have focused on their symbolic 'structure': the novice is seen as making a symbolic journey between two worlds, the world of children and the world of adults, and his journey is marked – at the boundary – by passage through a social 'no-man's land'.[10] But the fault, or rather the limitation, of this sort of structuralist analysis is that it fails to recognise that events which at one level have symbolic, non-literal, significance may at another level have practical consequences of quite another kind. Thus, to take examples from a different area, a structuralist analysis of incest taboos may show that they protect people from conceptual confusion between the cultural categories 'wife' and 'mother',[11] but it fails to recognise that they also protect people from the deleterious genetic consequences of inbreeding; or a structuralist analysis of the boiling of an egg may show that the egg is undergoing a culturally significant transition from the status of the 'raw' to that of the 'cooked',[12] but it fails to recognise that at the same time the egg is undergoing biochemical changes which render it more digestible.

Let us consider the case of the egg a bit more closely. The egg's change of social status depends on a cultural act, cooking: the egg is taken to a sacred place – the kitchen – and subjected to four minutes in a liminal 'no egg's land' – the pan of boiling water. But as a result of this transition rite it is not only the external, socially determined status of the egg that changes, its internal status changes too. It becomes – if the liminal period is long enough – hard-boiled, and the property of being hard-boiled is something private to the egg: it would remain hard-boiled if it were taken to the moon.

Now, surely the ritual initiation of a child may likewise bring about changes not only in the child's external relations to the social world but also in his private relations with himself. Rites of the kind I have described are not, after all, 'merely' symbolic: the child is frightened, hurt and mystified in earnest. Things are happening *in* him as well as *to* him.

One of the crucial things that must be happening in him is that he is internalising a new range of psychological concepts. The ritual trial provides for the child a forced introduction to feelings which were previously unknown. And so he gains a capacity for insight which may prove of inestimable value to his future attempts to interpret the behaviour of other human beings: insight into why people act the way they do in *non-ritual* situations where comparable feelings are aroused – in war, in the hunt, in oppression, in defeat, in the face of natural or supernatural terror.

The reader may now realise with dismay that in extolling the benefits of suffering, here as in the last chapter, I may seem to be providing an apology for cruelty. Given that a child, through suffering himself, gains insight into the behaviour of other people when in trouble, it follows that the agents of his suffering should be regarded as his teachers – and his benefactors. Being committed to the premise, I am some way committed to the conclusion.

Yet I should stress that I do not, as for example Nietzsche did, consider suffering and the infliction of suffering to be in their own right sources of strength and moral profit. Nietzsche's vision was black and gleeful:

> Examine the lives of the best and most fruitful men and peoples, and ask yourselves whether a tree, if it is to grow proudly into the sky, can do without bad weather and storms: whether unkindness and opposition from without . . . do not belong to the favouring circumstances without which a great increase in virtue is hardly possible.
>
> Almost everything we call 'higher culture' is based on the spiritualisation and intensification of cruelty – this is my proposition.[13]

Nor do I go along with the milder version of that philosophy proposed in a radio talk by A. S. Byatt: 'I think people need bad news, images of horror and disaster, and that they always have . . . I think the capacity to imagine disaster is a primitive and essential part of the human capacity to live long and survive.'[14]

Nietzsche tricks us with an empty metaphor. 'Ask yourselves whether a tree . . . can do without bad weather and storms.' The answer, if you think about it, is that a tree can do very well without bad weather and storms. But, and this is the point which Nietzsche missed, trees have no need to have insight into the sufferings of other trees. If a tree in the forest depended for its survival on its capacity to do introspective psychology, and if the other trees were themselves liable to be victims of bad weather and storms, then the first tree would indeed need to have shared their experience.

I do not believe for a moment that people grow more proudly into the sky because of suffering, but I do believe that they grow wiser as psychologists. In the course of their lives the members of any human social group meet with, and are challenged to understand, the suffering of others. 'Man that is born of woman', said the preacher, 'hath but a short time to live and is full of sorrow.' To understand man one must understand sorrow, and to understand sorrow one must have suffered it oneself.

Institutional fantasy: drama

During the Ndembu rite of circumcision the mothers of the novices 'sang softly and cried; they felt sad and afraid, not good at all'.[15]

In discussing initiation rites I have focused on the experience of the novice rather than on that of the spectator. It is the novice who is, as it were, guest of honour at the experiential feast, and it seems fair to assume that it is he who has most to gain in terms of psychological instruction. But there are clearly rich pickings to be had for the spectator too. Few people, I imagine, least of all a blood-relation, can stand by and watch an initiation rite unmoved. If it is true, as I've argued, that people can learn about human grief by witnessing the death of a pet dog, how much deeper must be the effect upon a mother who witnesses the symbolic rape and murder of her son.

In citing the mother's response as an example of 'spectator involvement' I must, however, allow that the circumstances are unusually favourable. First, it *is* her son: the personal relationship between spectator and actor could hardly be closer or more genuine. Second, the events she witnesses, although symbolic on one level, are on another level horrifyingly real.

Suppose, by contrast, that the relationship were illusory and the events were fake. Suppose that the mother were a Hampstead housewife, the novice a hireling from a theatrical agency, and the events no more than a dramatic fiction played out upon a London stage. Should we still expect the lady in the audience to feel sad and afraid, not good at all?

I doubt whether this particular play has yet been staged. But there are precedents enough to suggest that if it ever were the lady would be well advised to take her handkerchief along with her. Three thousand years of institutional drama have demonstrated the readiness of anonymous spectators to be roused to peaks of feeling by the spectacle of imaginary events upon the stage. Women are said to have gone into labour at the sight of Aeschylus' Furies; in a Chinese market town in 1678 a spectator leapt upon the stage and stabbed to death the actor who played the treacherous Ch'in Kuei; when the fictional Grace Archer (heroine of the radio soap-opera) 'died' in a fire in 1955 the Archer family, c/o the BBC, was deluged with wreaths and messages of sympathy.

Drama (in the theatre, in the cinema, on radio or television) is patently one of the most powerful techniques society has available for arousing – and so potentially expanding – the experience of its members. In our own civilisation drama has in fact now taken such a hold that for many people it provides a richer and more varied source of experience than real life. In pubs up and down Great Britain the conversation is as likely to turn on what happened in last night's television play as on what happened in the streets and homes of the local village. (I suspect it can only be the coyness of script-writers – or perhaps their fear of an infinite regress – which prevents the Archer family from discussing Coronation Street, while the denizens of Coronation Street discuss the Archers.)

But familiarity breeds contempt, and the very popularity of packaged drama has led some critics to be cynical about its benefits. I do not share their view. On the contrary, I believe that drama, in all its varied forms, represents an important cultural mechanism for introducing people to novel ways of feeling, supplementing their education as natural psychologists, and thus in the long run promoting mutual understanding between members of the social group. That it has come into its own in the context of modern civilisation is hardly surprising. In a culture where people might otherwise be condemned to exist, much of the time, like Matthew Prior's married couple –

> Without Love, Hatred, Joy or Fear,
> They led – a kind of – as it were;
> Nor Wish'd nor Car'd nor Laugh'd nor Cry'd:
> And so they liv'd: and so they dy'd

– it is just as well that we have plays, television and films to nourish our imagination, make us feel. When the local harvest fails, drama provides a Red Cross parcel for the mind.

Despite its importance and its prevalence, drama nevertheless remains a sophisticated art. In the earlier sections of this chapter I said nothing of the 'art' of pet-keeping or the 'art' of ill-treating a novice because, on the face of it, there is comparatively little to it. But the presentation of drama is a different matter: if the professional stage actor is to succeed in engaging the emotions of a strange spectator, he must have special skills.

Let me remind you of the two routes by which an actor can lead the spectator into new areas of subjective experience. (i) The actor may encourage the illusion of a personal relationship between the spectator and himself, so that the spectator becomes concerned about the actor's (apparent) fate. (ii) The actor may present himself as an attractive model, so that the spectator is tempted to identify with or imitate him (for example in a dream).

Yet, before the actor can get anywhere with the spectator, it is essential that his act should be convincing. Although the spectator may be prepared to make certain allowances for elements of falsehood in the staging of a play, he will not accept from the actor the false portrayal of character or sentiment. F. L. Lucas was right that 'only the more foolish type of public wants live rabbits in *A Midsummer Night's Dream*',[16] but equally only the more foolish type of public (or those who have been reading too much Brecht) wants un-lifelike 'alienated' heroines and heroes. The rabbits may be made of cardboard, but Hermia's expression of hatred for Helena must ring true.

But how shall that be done? The woman who plays Hermia, like as not, has no strong feelings either way for the woman who plays Helena:

> Is it not monstrous that this player here,
> But in a fiction, in a dream of passion,
> Could force her soul so to her own conceit
> That from her working all her visage wann'd,
> Tears in her eyes, distraction in her aspect,
> A broken voice, and her whole function suiting
> With forms to her conceit? And all for nothing!
> For Helena!
> What's Helena to her or she to Helena
> That she should spit at her?

Stanislavsky, with his famous 'method', gave a straight answer. The woman who plays Hermia must so far believe in her part that, leaving behind her life outside the theatre, she becomes reincarnated on stage in the form of the young Athenian girl whose lover has just been stolen by her best friend Helena. 'An actor must above all believe in what is happening around him and in what he himself is doing . . . he must live his part inwardly, and then give to this experience an external embodiment.'

In urging that an actor must live his part inwardly Stanislavsky was in effect demanding that the actor himself should excel as an introspective psychologist. I have talked all along of the task of the psychologist as being to create, on the basis of his own experience, a mental model of the behaviour of another person. The method actor's task is all of that and more, for he must go one better and turn his mental model into flesh and blood. In *An Actor Prepares* and his other manifestos Stanislavsky's advice reads, in effect, like a primer of natural psychology:

Inner characterisation can be shaped only from an actor's own inner elements. These must be felt and chosen to fit the image of the character to be portrayed . . . If this is effectively prepared the outer characterisation should naturally follow . . . Let every actor achieve this outer characterisation by using material from his own life, from that of others, real or imaginary, by using his intuition, self-observation . . . Do you expect an actor to invent all sorts of new sensations, or even a new soul, for every part he plays? How many souls would he be obliged to house? Can he tear out his own soul and replace it by one he has rented as being more suitable to a certain part? Where can he get one? You can borrow *things* of all sorts, but you cannot take feelings from another person. You can understand a part, sympathise with the person portrayed, and put yourself in his place, so that you will act as he would. That will arouse feelings in the actor that are analogous to those required for the part . . . Analysis studies the external circumstances and events in the life of a human spirit in the part; it searches in the actor's own soul for emotions common to the role and to himself, for sensations, experiences, for any elements promoting ties between him and his part.[17]

The actor, then, must learn to cultivate and use his intuition. The Moscow Arts Theatre, the Arts Studio in New York, the Royal Academy of Dramatic Art in London – in subjecting their students to a daily round of fantasy, imitation, teasing, childish play – may well have a better claim than any University Department to be true schools of psychology.

But the actor, as an actor, has other serious work to do. His job on stage is not just to explore his own possibilities of feeling but to kindle new feelings in an audience: having been a pupil, he must become in his turn a teacher of psychology. And for that he must do more than live his part convincingly, he must make the character he plays *engaging*.

Why should a spectator mind about what happens to a character on stage? Why should he – mistakenly – assume that the character's fate has anything to do with him?

The illusion of personal relationship between the actor and spectator is typically fostered in two different ways: first, by tricks of theatrical direction, plot and staging (what Erving Goffman calls 'dramatic frame');[18] second, by the personality of the individual actor.

Dramatic frame

In real life the people we have close relationships to are distinguished from strangers in several obvious ways: we know, for example, an unusual amount about their private lives, we expect them to tell us things which they would not reveal to others, we accompany them at moments of crisis, we enter their private homes and meet their families and friends, we expect them – and permit them – to drop their guard in front of us in matters of dress, manners and decorum.

By these criteria, the dramatic hero is – so it seems – far from being a stranger to the audience. On stage Hamlet openly discusses suicide with us, Juliet tells us the secrets of her bedchamber, we are present in the drawing room of the Doll's House when Nora and her husband quarrel, Antigone looks despairingly to us for help as the guards close in around her. On screen we are drawn in closer still by an illusion of physical intimacy: the heroine's face fills our field of view as she gazes lovingly into our eyes, we look over the mother's shoulder as she holds the baby to her breast, the villain aims a blow straight at our jaw.

That a spectator, in these circumstances, should half-imagine himself personally related to the actor is wholly understandable. The 'fallacy' which leads him on, known to logicians as 'affirming the consequent', is one which has a powerful hold on human reasoning. 'If *p* is the case then *q* will be the case; *q* is the case; so *p* must be the case' – 'If I am this man's friend he will tell me his trouble; he is telling me his trouble; so I must be his friend'; 'If I am this woman's lover she will show me her naked body; she is showing me her naked body; so I must be her lover.'

The spectator's reasoning is not of course quite so simple as all that. But that is the beginning of it, and the follow-through from half- to wholly imagining the spurious relationship is aided by another feature of the drama: its location in the temple-like setting of the theatre or the cinema, where the spectator is encouraged to let go his genuine relationships and leave them, like his slippers, at the door. He is ushered into the sanctum of the theatre by uniformed attendants, the lights are dimmed, music plays to mark the boundary between reality and fantasy, and for the next two hours he sits in total darkness mutely watching the spectacle. As a means of temporarily 'depersonalising' the spectator, of making him forget himself, this setting is ideal.

And if the illusion is to work the spectator must indeed forget himself – forget his *previous* self. For it takes two people to make a relationship, and correspondingly two imaginary people to make an imagined relationship: not only must the spectator be wrong about the actor, he must be wrong about himself.

Perhaps it is the very lack of the conditions which make for such depersonalisation which prevents television or radio drama from being so powerful as that on stage or screen. When a spectator, instead of having to go to the drama-house, can sit comfortably at home, the degree to which he will be persuaded to surrender his own personhood is certain to be much reduced. It may not be too difficult for the 'home viewer' to feel himself related to the keeper of the Ambridge pub, but only someone who enters the cavern of a theatre is likely to forget himself so far as to imagine himself a brother to the Prince of Denmark.

The actor's personality

While the conventions of production and staging go a long way towards explaining the power that drama has to draw together the actor and spectator, the aura that surrounds an individual actor can also add substantially to the illusion. 'There are certain actors who have only to step on the stage and the public is already enthralled by them,' said Stanislavsky: 'What is the basis of the fascination they exercise? It is an indefinable, intangible quality . . .'.

Intangible, certainly, but perhaps not indefinable. The quality in question, which Stanislavski called stage charm, manifests itself more generally in the power of any 'charismatic' figure to engage the attention and devotion of an audience. It rests, I believe, on the capacity of such a figure to elicit even from a total stranger a feeling of *déjà connu* – the sense of a preformed relationship.

The relationship is indeed preformed, but not between the spectator and the actor. It was formed, most probably in childhood, between the spectator and some valid object of affection – a parent, sibling, teacher, childhood friend. The partner of the earlier relationship has vanished, but the valent energy of the relationship survives, a silver hook looking for a golden eye. The charismatic figure seemingly provides for every hook an eye.

What gives that kind of power? If I were to say that it has to do with an ambiguous or enigmatic presentation of the self I might seem to be saying little. But perhaps I shall make myself clearer by quoting a description of one person's reaction to the enigma of the Mona Lisa, the most charismatic portrait ever painted. The words are Walter Pater's:

> Hers is the head upon which all 'the ends of the world are come' . . . All the thoughts and experience of the world are etched and moulded there . . . the animalism of Greece, the lust of Rome, the reverie of the middle age with its spiritual ambition and imaginative loves, the return of the Pagan world, the sins of the Borgias. She is older than the rocks among which she sits; like the vampire, she has been dead many times, and learned the secrets of the grave; and has been a diver in deep seas, and keeps their fallen days about her; and trafficked for strange webs with Eastern merchants; and, as Leda, was the mother of Helen of Troy, and, as Saint Anne, the mother of Mary; and all this has been to her but as the sound of lyres and flutes, and lives only in the delicacy with which it has moulded the changing lineaments and tinged the eyelids and the hands.[19]

The Mona Lisa's charm lies in that very sorcery which allows her to appear to the spectator at one and the same time as Leda, St Anne or whom else he cares to see in her. Leonardo, in making a portrait of a Florentine merchant's wife, succeeded in painting a picture of a woman who is in the truest sense *déjà connu*, a woman whom everyone knows, is related to, hooks into, at the first encounter.

The greatest charismatic actors – Charles Chaplin, Greta Garbo, Marilyn Monroe – must have some of the same quality. By face and manner they hint at possible existences, yet never finally confirm or deny them. And so they invite the spectator to make his own interpretation, to draw boundaries around a persona which is essentially unbounded. He is free to project upon them his past loves and fears – to see in them his wife, his nurse, his friend, his other self – in a way which any more ordinary, concrete face or personality would disallow. The face of the Mona Lisa – and the face of Garbo – provide, as it were, a human wishing-well.

*

I began this section on drama with the example of a woman's response to the circumcision of her son. In her case her relationship to the son was all too genuine and the events played out before her all too real: and so, poor thing, she sang softly and cried. Let me end with the example of the response of a spectator at the cinema to Anna Karenina's suicide. In his case his relationship to Anna (or to Garbo, who is playing her) is anything but genuine and the events anything but real: yet he, in his turn, goes wet-eyed from the cinema.

The cinema-goer weeps (i) because Garbo, the trained actress, has convinced him that she is Anna, (ii) because this Anna, as scripted and directed by the film-maker, has trustingly revealed herself to him as to a friend, (iii) because this Anna, as played by Garbo, has a face that he already dimly recognises as representing the idealised lover–mother–child, (iv) because in the darkness of the cinema he has forgotten the true limits to what he is himself. It amounts, in all, to a remarkable piece of 'experiential engineering' by the film industry.

But does not a piece of engineering need an engineer? Does not an engineer need a client? And must not both of them know *what* they're doing? Does it really make sense to suppose Sam Goldwyn – or Pedigree Pet Foods Ltd or the Iatmul *tumbuan* – are in business to give people greater insight into the behaviour of their fellow human beings? Few people, surely, can be said to go to the theatre or cinema with the express intention of broadening their psychological vocabularies; fewer people can be said, for *that* reason, to keep pets; and fewer still, for *that* reason, to submit themselves to ritual circumcision.

My answer – which I have no doubt can and will be misconstrued – is that it really doesn't matter what the participants in these ceremonies themselves may think to be the purpose of their actions: what matters are the consequences, all of them, in practice. The spectator of *Anna Karenina*, who has sympathised with Anna, pitied her, foreseen the coming tragedy, and watched helplessly as her body was crushed beneath the train, the spectator who has *by that fact* gained greater insight into himself and other people, has increased his fitness both as an individual and as a member of society. Likewise with the other cultural mechanisms that I have been discussing. Likewise, too, with the natural mechanisms discussed in the last chapter.

I do not for a moment mean to say that this is *all* there is either to the cultural or to the natural mechanisms, any more than a sociobiologist would say, for example, that the avoidance of inbreeding is all there is to the incest taboo. Biologists – perhaps rather more than cultural anthropologists – are used to the idea that a specific trait may in fact have several different 'beneficial consequences' for an animal.[20] The feathers of a bird provide it with thermal insulation, they provide it with a flight mechanism, and they give it a pigmented surface for display. Take away a robin redbreast's feathers and the bird may die of cold, *or* of flightlessness *or* of no longer being a redbreast. At different times natural selection will act more strongly on one consequence than on another, but they all contribute to the robin's biological survival. So too, within the context of a culture, a film may serve to provide emotional catharsis, to line the pockets of the film-maker, or simply to keep people off the streets at night, as well as contributing to the spectator's education. The show goes on for different reasons. But even if only one of the benefits of people's subscribing to a cultural institution is that it increases their insight into other human beings, then almost certainly that is one of the most important factors in keeping the institution alive.

AESTHETICS AND ILLUSION

When the question is whether a thing be beautiful, we do not want to know whether anything depends or can depend, for us or for anyone else, on the existence of the object, but only how we estimate it in mere contemplation . . . If nature had produced her forms for our satisfaction it would be a grace done to us by nature, whereas in fact we confer one upon her.

Immanuel Kant, *Critique of Judgement*, 1790

9

THE ILLUSION OF BEAUTY

'Beauty is truth, truth beauty' to the poet. But a biologist is bound to regard beauty – at least man-made beauty – as something closer to a lie. A lie, admittedly, of a peculiar kind, but of a kind to which both men and animals are specially vulnerable.

If I give a hungry dog a solution of saccharine it will lap it up; if I show a cock robin a bundle of feathers with a red patch on its underside the robin will attack it; and if I show a man an abstract painting or play him a piece of music he will, if he thinks it beautiful, stop to watch or listen. There is, I believe, a formal similarity between these cases. In each we have an animal making a potentially useful and relevant reponse towards an *inappropriate* sensory stimulus. But there is also a rather basic difference, namely that in the first two cases we have a good scientific explanation of what is going on, while in the third we're almost ignorant. With the saccharine and the red-breasted bundle of feathers we know what the artificial, 'illusory', stimulus corresponds to in nature and we know how the dog's or the robin's behaviour would in normal circumstances contribute to its biological survival. Saccharine tastes like sugar and it is biologically adaptive for a hungry dog to eat sugar; the bundle of feathers looks like another male robin and it is biologically adaptive for a robin to drive an intruder from his territory. But in the case of man's response to a beautiful work of art we have no clear idea either of what the work of art corresponds to in nature, or of why it should be biologically adaptive for men to *like* the natural counterpart (whatever it may be).

It is with these fundamental questions in the biology of aesthetics that this chapter is concerned. I plan first to try to define the particular quality which things of beauty have in common and then to suggest a possible reason why men – and, for that matter, animals – should be attracted to the presence of that quality.

Seventy years before Darwin published *The Origin of Species* Thomas Reid suggested how a modern biologist might proceed:

By a careful examination of the objects to which Nature hath given this amiable quality [of Beauty], we may perhaps discover some real excellence in the object, or at least some valuable purpose that is served by the effect which it produces upon us. This instinctive sense of beauty, in different species of animals, may differ as much as the external sense of taste, and in each species be adapted to its manner of life.[1]

Yet it is easy to dismiss Reid's manifesto. The injunction to 'examine carefully' the objects of beauty would be fine were it true that different individuals of the same species did find the *same* objects beautiful. But one of the central problems of aesthetics has always been that, in man at least, there is no clear consensus. The point was forcefully made by Maureen Duffy in her review of Jane Goodall's book *In the Shadow of Man*. Jane Goodall had written 'But what if a chimpanzee wept tears when he heard Bach thundering from a cathedral organ?', to which Miss Duffy replied 'What indeed if an Amazon pigmy or a 19th century factory hand wept tears at such a minority western cultural phenomenon?'

The way out for some critics, confronted with the diversity of individual taste, has been to react with the cynicism of Clive Bell, stating that 'any system of aesthetics which pretends to be based on some objective truth is so palpably ridiculous as not to be worth discussing'.[2] But William Empson scorned such anti-rationality: 'Critics', he wrote, 'are of two sorts: those who merely relieve themselves against the flower of beauty, and those, less continent, who afterwards scratch it up. I myself, I must confess, aspire to the second of these classes; unexplained beauty arouses an irritation in me . . .'.[3] My own sympathies lie, of course, with Empson. But the roots of the flower of beauty may go deep and if we are to expose them undamaged we should start scratching at some distance from the stem.

The problem of looking for common principles behind apparent diversity is not peculiar to aesthetics. Very similar problems have arisen in other disciplines, notably in linguistics and in anthropology. The breakthrough in these fields came through applying the methods of *structuralism*. I believe that a structuralist approach is the key to a science of aesthetics.

In his discussion of the analysis of myth, Claude Lévi-Strauss wrote as follows:

For the contradiction which we face is very like that which in earlier times brought considerable worry to the first philosophers concerned with linguistic problems . . . Ancient philosophers did notice that certain sequences of sounds were associated with definite meanings, and they earnestly aimed at discovering the reason for the linkage between *these* sounds and *that* meaning. Their attempt however was thwarted from the beginning by the fact that the same sounds were equally present in other languages although the meaning they conveyed was entirely different. The contradiction was surmounted only by the discovery that it is the combination of sounds, not the sounds themselves, which provides the significant data.[4]

He went on: 'If there is a meaning to be found in mythology, it cannot reside in the isolated elements which enter into the composition of a myth, but only in the way those elements are combined.'

Following this lead, it would seem fruitful to search for the essence of beauty in the *relations* formed between the perceived elements. As it happens, just such an approach was proposed in 1808 by the philosopher Herbart: 'The conclusion is that *each* element of the approved or distasteful whole is, in isolation, indifferent; in a word, the *material* is indifferent, but the *form* comes under the aesthetic judgement . . . Those judgements which are commonly conceived under the name of taste are the result of the perfect apprehension of relations formed by a complexity of elements.'[5]

But it is one thing to point to the importance of relations, another to say *what* relations are important, and another still to say *why*.

Lévi-Strauss himself, in so far as he had anything to say about aesthetics, tended to regard works of art merely as a special sort of myth. For him the work of art is a 'system of signs' which conveys a message. To understand the message we must make an equation between the *relations* among the signs and the *relations* among the things signified.

No doubt such myth-like works of art exist. We know for instance of a Chinese scholar, Lyng Lun, who 2,500 years before Christ strung together five tones of oriental music, explained them, formed them into a system, and gave them strange names, every tone being called after a social stratum from the emperor down to the peasant: *kong*, the emperor; *chang*, the minister; *kyo*, the burgher; *tchi*, the official; *yu*, the peasant.[6] Within such a system almost any piece of music must, if interpreted in a structural way, have carried a potential social message. In the field of graphic art, Caroline Humphrey has shown how the magical drawings of the Mongolian Buryat people embody structuralist devices which make the drawings effectively into 'visual texts'.[7] And almost certainly similar sign-systems are at work within the mainstream of Western painting. Christopher McManus and I found evidence that Rembrandt, for instance, may have made use of a simple sign-system in his painted portraits, whereby the social status of the subject of the portrait was indicated by the left or right turn of his head (see Chapter 10).

But be that as it may, these sign-systems, where they exist, serve primarily a semantic function, not an aesthetic one. They do not lend *beauty* to a work of art. If structuralism is to help in pointing to relations which are *aesthetically* satisfying it must take a different turn.

Few people have written with more insight about beauty than the poet Gerard Manley Hopkins. Hopkins is hardly to be called a 'structuralist' since the name had still to be invented in his lifetime, yet not only did he see that the essence of beauty lies in certain relations but he attempted explicitly to define what those relations are. In 1865 he wrote a paper for his tutor at Oxford in the form of a 'Platonic dialogue' between a student and a professor in a college

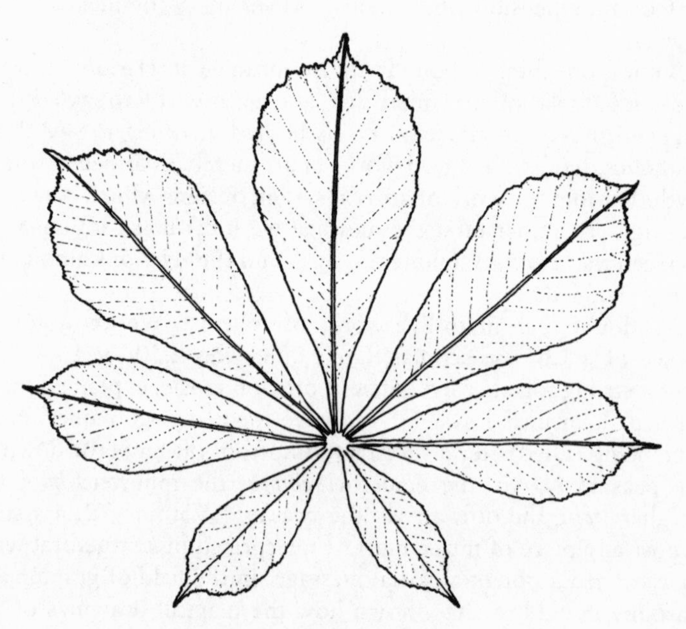

Figure 4. The chestnut-fan.

garden.[8] The two of them fall to discussing the beauty of the garden and they dwell in particular on the leaves of a chestnut tree. The professor holds forth on the structural relations within the chestnut-fan (Figure 4), pointing out how each leaf is a variation with a difference of the common pattern, how the overall shape of the fan shows mirror symmetry, the left half being a perfect reflection of the right, while in other ways the internal reflections are tantalisingly irregular – each of the large oblique leaves, for instance, being reflected by an exact copy of itself in miniature; and he discusses, too, the relation between the leaves of the chestnut and the leaves of other trees, drawing attention to the way in which the chestnut leaf, being fatter at the outer than the central end, has a shape opposite to the common shape shown, say, by the leaf of an elm. The professor continues:

> 'Then the beauty of the oak and the chestnut-fan and the sky is a mixture of likeness and difference or agreement and disagreement or consistency and variety or symmetry and change.'
>
> 'It seems so, yes.'
>
> 'And if we did not feel the likeness we should not feel them so beautiful, or if we did not feel the difference we should not feel them so beautiful. The beauty we find is from the comparison we make of the things with themselves, seeing their likeness and difference, is it not?'

Before long they move on to the subject of poetry:

> 'Rhythm therefore is likeness tempered with difference . . . And the beauty of rhythm is traced to the same causes as that of the chestnut-fan, is it not so?' . . .
>
> 'What is rhyme? Is it not an agreement of sound – with a slight disagreement?' . . .
>
> 'In fact it seems to me that rhyme is the epitome of our principle. All beauty may by a metaphor be called rhyme, may it not?'

In 1909 Christiansen coined the word 'Differenzqualität' to refer to what Hopkins had called 'likeness tempered with difference'. And shortly afterwards the writers of the school of Russian formalism propounded a system of aesthetics based on essentially similar structuralist ideas. In England the philosopher A. N. Whitehead wrote of rhythm: 'The essence of rhythm is the fusion of sameness and novelty; so that the whole never loses the essential unity of the pattern, while the parts exhibit the contrast arising from the novelty of their detail. A mere recurrence kills rhythm as does a mere confusion of differences. A crystal lacks rhythm from excessive pattern, while a fog is unrhythmic in that it exhibits a patternless confusion of detail.'[9]

Here then we have the beginnings of an answer as to what relations lie at the heart of beauty. 'All beauty may by a metaphor be called rhyme.' What is rhyme like? Well, let us have an example:

cat rhymes with *mat*;
cat does not rhyme with *table*;
cat does not rhyme with *cat*.

Taking rhyme as the paradigm of beauty, let me turn at once to the fundamental question: Why do we *like* the relation which rhyme epitomises? What is the biological advantage of seeking out rhyming elements in the environment?

The answer I propose is this: Considered as a biological phenomenon, aesthetic preferences stem from a predisposition among animals and men to seek out experiences through which they may *learn to classify* the objects in the world about them. Beautiful 'structures' in nature or in art are those which facilitate the task of classification by presenting evidence of the 'taxonomic' relations between things in a way which is informative and easy to grasp.

Three steps are needed to justify this argument. First, an explanation of why classification should be important to biological survival. Second, an explanation of why particular structures such as those exemplified by rhyme should be the best way of presenting material for classification. Third, evidence that men and animals have a propensity to classify things and that they are attracted in particular to the presence of rhyme.

Why is classification important?

In order to be effective agents in the natural world, animals require the guidance of a 'world model', an internal representation of what the world is like and how it works. This model enables them to predict in advance the characteristics of 'recognisable' objects, to anticipate the likely course of events in the environment, and to plan their behaviour accordingly. The role of classification in this context is to help organise sensory experience and to introduce an essential economy into the description of the world. An effective classification system is one which divides up the objects in the world into discrete categories according to criteria which make an object's membership of any particular class a relevant datum for guiding behaviour: the objects in any one class may differ in detail but they should share certain essential features which give them a common significance for the animal. Such a classification system will reduce the 'thought load' on the animal, expedite new learning, and allow rapid and efficient extrapolation from one set of circumstances to another.

We may be sure that any animal which could not or did not classify things effectively – which could not recognise the likenesses between things – would not have a chance of surviving for long. And so, in the course of evolution, there must have been very strong pressures on animals to perfect techniques of classification, on a par perhaps with those that have made eating and sex evolve into such efficient and dominant activities. I shall argue that, just as with eating or sex, an activity as vital as classification was bound to evolve to be a *source of pleasure* to the animal. Both animals and men can, after all, be relied on to do best what they enjoy doing.

But I am anticipating. The next step of the argument is to demonstrate the relevance of rhyme.

On what kind of 'evidence' are classification systems based?

The young animal's task of imposing a system of categories upon the world is comparable to that which faces a zoological taxonomist when he sets out to classify the animal kingdom. We may assume that the goal before the animal is in some sense 'given', that he has an innate predisposition to develop *a* system of categories, but that the actual system he arrives at must be largely based upon his own experience. How does the animal – and the zoologist – proceed? I would suggest he works through the following stages:

(i) he makes a preliminary reconnaissance and from this forms certain hunches about how his world is constituted, what kinds of classes of objects it contains and what are the distinguishing criteria;

(ii) he seeks further evidence to test the 'validity' of these criteria and at the same time to acquaint himself with the diversity which may exist *within* each class;

(iii) to the extent that his criteria prove successful he adopts them as permanent guidelines for future classification, while to the extent that they fail he abandons or revises them.

Successful criteria will on the whole be those which yield a system of classification which is at once *unambiguous* (that is, objects belong to one class only); *exhaustive* (that is, every object belongs to some class or another); and *useful* (that is, objects in the same class may be treated for some practical purpose as identical). Thus a zoologist who comes up with a simple classification of animals which divides them into two classes, vertebrates and invertebrates, has produced a scheme which meets all three requirements. But a division of animals into meat-eaters and plant-eaters, for example, fails to be unambiguous since there are some animals which eat both; a division into swimming animals and flying animals fails to be exhaustive since there are some animals which neither swim nor fly; a division into zoo-animals and wild animals fails to be useful since there is no purpose (for a zoologist) in treating the members of either of these classes as identical.

In order to examine the process of seeking evidence to test the criteria for distinguishing between classes, let me continue with a zoological example. Imagine that the taxonomist is concerned to classify warm-blooded vertebrates. In making a preliminary survey he meets a cat, a dog and a hen, and he notices that the cat and the dog are covered with hair, but the hen is covered with feathers. On this basis he sets up two putative classes, mammals and birds, defined respectively as animals which have hair and animals which have feathers. His next step is to look for further examples to test his ideas. Suppose that the next animal he meets is a horse, and the one after that a rabbit. Applying his criteria he discovers that these animals fit neatly into the category of mammals. Then perhaps he meets a sparrow, then a mouse, then a parrot, and he is pleased to find that while the mouse is clearly a mammal the sparrow and the parrot fit the definition of a bird. Looking further he meets another cat, but on this occasion he pays it little attention since it tells him nothing new. And later on he meets an octopus, but since this is not a warm-blooded vertebrate it can provide no evidence either way and again he shows no interest in it. Slowly, by accumulating evidence, he establishes that his criteria do indeed serve to make unambiguous distinctions, and at the same time he becomes familiar with the range of different animals that fall within each class. It remains of course for him to show that his classification is a useful one – that it serves some purpose to group mice and horses or hens and parrots together.

Certain principles of how to gather evidence emerge. The zoologist needs to prove that his criteria serve both to *group* different animals together and to *separate* one group from another. Accordingly he looks for two kinds of examples: (i) sets of animals which share a particular distinctive feature, and (ii) other sets of animals which share a contrasting feature. Thus he looks in effect for 'likeness tempered with difference' – 'rhyme' – and for *contrast* between sets of rhyming elements. But he is not interested in seeing repetitive examples of the same animal, nor in seeing an animal which is altogether different from the others and thus lies beyond the scope of his classification – 'a mere recurrence kills rhyme, as does a mere confusion of differences'.

To pursue this metaphor of the taxonomic 'poem':

horse rhymes with dog;
hen rhymes with parrot;
horse and dog contrast with hen and parrot;
horse does not rhyme with horse, nor hen with hen;
neither horse nor dog nor hen nor parrot rhyme or contrast in a relevant way with octopus.

Now to the nub of my argument. I believe that the same principles which apply to the zoological taxonomist apply to every animal who needs to classify the world about him. If it is helpful for the taxonomist to look for 'rhymes' in his materials, so it is helpful for the animal to do so. It is for this reason that we have evolved to respond to the relation of beauty which rhyme epitomises. At one level we take pleasure in the abstract structure of rhyme as a model of well-presented evidence, and at another we delight in particular examples of rhyme as sources of new insight into how things are related and divided.

Let me move on to the next stage of the argument and give evidence that men and animals do indeed take pleasure in classifying things and, on that account, are especially attracted to rhyme.

The propensity to classify – and the love of 'rhyme'

'Learning', said Aristotle, 'is very agreeable, not only to philosophers but also to other men.'[10] What evidence is there that classification – the core of learning – is agreeable to men and to animals also?

For experimental evidence of a general kind we may look to the many studies of exploratory behaviour. Comparative psychologists have found that, in almost every species studied, animals will work to be exposed to novel sensory stimuli. Indeed, 'stimulus novelty' is the most universal reinforcer of behaviour which is known. In my own work with monkeys I have found that monkeys will even work to look at abstract paintings, and prefer such pictures to pictures of appetising, but familiar, food. Recent experiments strongly suggest that when monkeys work to look at pictures they do so because the picture presents them with a challenge to incorporate new material into their model of the world: pictures of familiar objects hold their attention far less long than pictures of objects for which they have no readily available category.[11] But while they do not spend long on thoroughly familiar things, neither, I should say, are they interested in looking at a total jumble. And that leads me on to the question of rhyme.

The significance of rhyme was in fact recognised by experimental psychologists some time ago, though they called it – and still call it – by the cumbersome name of 'stimulus discrepancy'. In the early 1950s a theory was propounded called the 'discrepancy theory', the gist of which is that men who have been exposed for some time to a particular sensory stimulus respond with pleasure to minor variations from that stimulus.[12] And confirmatory evidence has come from a number of studies. For instance, human babies who have been made familiar with a particular 'abstract' visual pattern take pleasure in seeing new patterns which are minor transformations of the original.[13] Among animals, it has been shown, for instance, that chicks who have been 'imprinted' early in life on an artificial stimulus soon come to prefer new stimuli which are slightly different from the one they are familiar with.[14] Neither babies nor chicks are attracted to stimuli which are wholly unrelated to what they have already seen.

I have pursued my own research with monkeys along these lines. But this is not the place to report the details of experiments. And it is not in fact to experimental evidence that I want to give most weight in this discussion. For there is much in the evidence of anecdote and common experience to substantiate the view that men, at least, take pleasure in one form or another of classificatory activity.

As we might expect, the tendency is most pronounced in children. Children have a thirst to know 'what things are'. They love especially to learn *names*, and to prove the power of their vocabulary with new examples. Picture books for children often serve no other purpose than as practical exercises in classification. The same animals – rabbits, hens, pigs – appear in the pictures again and again. 'Where's the bunny?' asks the child's mother, and with a smile of pleasure the child points a finger to yet another rabbit which rhymes with those he has already seen. The ability to name becomes tangible evidence of the ability to classify, and when the name for an object is not available children will often invent their own. The poet Richard Wilbur tells this story:

'I took my three-year-old son for a walk in the Lincoln woods. As we went along I identified what trees and plants I could . . . After a while we came to a stretch of woods-floor thick with those three-inch evergreen plants one sees everywhere in New England woods, and I was obliged to confess I didn't know what to call them. My three-year-old stepped promptly into the breach. "They're millows," he told me, "Look at all the millows." No hesitation; no bravado; with a serene Adamite confidence he had found a name for something nameless and brought it under our verbal control. Millows they were.'[15]

Yet while children may manifest the tendency most clearly, adult men often show an equally innocent delight in classifying, not least in naming. A poem by Robert Bridges called 'The Idle Flowers' mentions 83 different flowers by name in a poem only 84 lines long!

I have sown upon the fields
Eyebright and Pimpernel,
And Pansy and Poppy-seed
Ripen'd and scatter'd well.

And silver Lady-smock
The meads with light to fill,
Cowslip and Buttercup,
Daisy and Daffodil;

King-cup and Fleur-de-lys
Upon the marsh to meet
With Comfrey, Watermint,
Loose-strife and Meadowsweet;

And all along the stream
My care hath not forgot
Crowfoot's white galaxy
And love's Forget-me-not . . .

The reverse of the coin is the ridicule that is heaped on people who make mistakes with names. A. P. Herbert tells a story against himself, again to do with flowers:

'The anemias are wonderful,' I said. My companion gave me a doubtful glance but said nothing. We walked on beside the herbaceous border. 'And those arthritis, always so divine at this time of the year.' Again the dubious glance, and again no utterance except an appreciative 'Um'. I came to the conclusion that the young lady knew no more about flowers than I do.[16]

The concern with naming, carried to such an extreme in Bridges's poem, finds echoes in another remarkable aspect of human behaviour – the passion for *collecting*. Surprisingly there is only one student of human behaviour I know of who has thought collecting worthy of comment. In an essay called 'The Reflex of Purpose', Ivan Pavlov characterised collecting as 'the aspiration to gather together the parts or units of a great whole or of an enormous classification, usually unattainable'.[17] He went on:

If we consider collecting in all its variations, it is impossible not to be struck with the fact that on account of this passion there are accumulated often completely trivial and worthless things, which represent absolutely no value from any point of view other than the gratification of the propensity to collect. Notwithstanding the worthlessness of the goal, everyone knows the dedication, the single-mindedness with which the collector achieves his purpose. He may become a laughing-stock, a butt of ridicule, he may suppress his fundamental needs, all for the sake of his collection.

Yet while collectors are not usually credited with aesthetic sensitivity, their activity may not be far removed from other forms of aesthetic appreciation. Consider the nature of a typical collection, say a stamp-collection. Postage stamps are, in structuralist terms, like man-made flowers: they are divided into 'species', of which the distinctive feature is the country of origin, while within each species there exists tantalising variation. The stamp-collector sets to work to classify them. He arranges his stamps in an album, a page for the species of each country. The stamps on each page 'rhyme' with each other, and contrast with those on other pages.

But Pavlov was right: stamp-collecting is apparently a worthless activity. As we have moved through my examples, from an infant animal learning to recognise the objects in the world about him, to a child learning to name pictures in a book, to a man sticking stamps in an album, we have moved further and further from activities which have any obvious biological function. They are all, I submit, examples of the propensity to classify, but with each example the classification seems to have less and less direct survival value.

We should not be surprised. Earlier, I compared the pleasure men get from classification with the pleasure they get from sexual activity. Now, though sex has a clear biological function, it goes without saying that not every particular example of sexual activity needs to be biologically relevant to be enjoyable. Indeed, much normal human sexual activity takes place at times when the woman, for natural or artificial reasons, is most unlikely to conceive. And so too the process of classification may give pleasure in its own right even when divorced from its proper biological context. Once nature had set up men's brains the way she has, certain 'unintended' consequences followed – and we are in several ways the beneficiaries.

So let me turn, at last, to *beauty* – to examples of rhyme and contrast which people deem *aesthetically* attractive. I want first to consider not 'works of art' but certain natural phenomena which men call beautiful and yet which have no 'natural' value to us.

Among the wealth of examples of beauty in nature, I shall choose the case of flowers. Flowers have an almost universal appeal, to men of all cultures, all classes and all ages. We grow them in gardens, decorate our houses and our bodies with them, and above all value them as features of the natural landscape. They are regarded indeed as paragons of natural beauty, and I believe it is no accident that they are so admired, for in at least three ways flowers are the embodiment of 'visual rhyme'.

Consider first the static form of a simple flower such as a buttercup or daisy. The flower-head consists of a set of petals arranged in radial symmetry around a cluster of stamens, and the flower-head is carried on a stalk which bears a set of leaves. Petals, stamens and leaves form three sets of contrasting rhyming elements: each petal differs in detail from the other members of its class yet shares their distinctive shape and colour, and the same is true for the stamens and the leaves; the features that serve to unite each set serve at the same time to separate one set from another. Secondly, consider the flower's *kinetic* form. The living flower is in a continual state of growth, changing its form from day to day. The transformations which occur as the flower buds, blossoms and decays give rise to a temporal structure in which each successive form rhymes with the preceding one. Thirdly, consider groups of flowers. Typically each flowering plant bears several blooms, and plants of the same species tend to grow in close proximity, so that we are presented with a variety of related blooms on show together. But, more than this, groups of flowers of *different* species commonly grow alongside one another – daisies and buttercups beside each other in the field, violets and primroses together in the hedgerow. Thus while the flowers of one species rhyme with each other the rhyme is given added poignancy by the contrasting rhymes of different species. It is this last aspect that perhaps more than anything makes flowers so special to us. The flowers of different species are of necessity perceptually distinct in colour, form and smell in order that they may command the loyalty of pollinating insects. Men neither eat their pollen nor collect their nectar, yet flowers provide us with a kind of nourishment – food for our minds, ideally suited to satisfy our hunger for classification.

But flowers have no monopoly of natural beauty. In fact, almost wherever we come across organic forms we discover the structure of visual rhyme. Long before architects invented the *module*, nature employed a similar design principle, basing her living creations on the principle of replication – at one level, replication of structural elements within a single body; at another, replication of the body of the organism as a whole. But, at either level, the replicas are seldom, if ever, perfect copies: in the leaves of a tree, the spots of a leopard, the bodies of a flight of geese, we are presented with sets of 'variations on a theme'. And it is not only among living things that we find such structures, for inanimate objects too tend to be shaped by physical forces into 'modular' forms – mountain-peaks, pebbles on a beach, clouds, raindrops, ocean waves – each alike but different from the others. Thus, through its varied but coherent structure, a natural landscape can match the rhythmic beauty of a gothic church or of a musical symphony.

It is fundamental to my argument that people may find beauty in many different guises. Before I turn to art as such, let me say something about intellectual beauty, the beauty people find in academic scholarship. Pure science is, for most of its practitioners, an eminently aesthetic activity. The scientist's aim is to impose a new order on natural phenomena by uniting seemingly unrelated events under a common law. Artists have often misunderstood the nature of science. Keats, the poet, complained that Newton had 'unweaved the rainbow'. But Newton's achievement was near enough itself to poetry. He showed how the rainbow rhymed with the solar spectrum he cast with a prism on his study wall.

At an extreme among scholars, mathematicians may find their own kind of beauty in the relations between abstract numerical ideas. We non-mathematicians may sometimes catch the flavour of their abstract structures when we are shown the magical properties of certain ordinary numbers. I remember when as an eight-year-old child I was introduced by my grandfather to the number 142857. If this number is multiplied first by 1, then by 2, then by 3, and so on seven times, the following series is generated:

142857
285714
428571
571428
714285
857142
999999

Six rhyming numbers, and then the sudden unexpected contrast. Imagine my awe when ten years later I came across a proof that this is the only number which has such properties.

Children, monkeys, gardeners, stamp-collectors, mathematicians – all, I think, are engaged in similar aesthetic enterprises. 'Obscurity', Hume wrote, 'is painful to the mind as well as to the eye', and it should come as no surprise to know that a late Professor of Formal Logic at Cambridge University was also a prodigious collector of stamps and butterflies. But he was not, it must be said, an artist. Where does art fit in to this account of beauty?

Until the beginning of this century most paintings were only half concerned with beauty, their other role being to tell a story by means of representation, expression, symbolism and so on. Only with the advent of pure abstractionism did the goal of some artists come to be the creation of great works which were 'merely' beautiful. If we consider the finest examples of modern abstract art, for example the works of Vasarely and of Calder, it is I think easy, perhaps too easy, to see how their structure is essentially that of a visual poem built up on the basis of rhyme and contrast between visual elements. Recently some artists have returned to the use of representational elements as the material for creating purely abstract structures. Suzi Gablik's painted collages – at first sight a crazy scrapbook of animal images (see Plate 1) – are a bold attempt in this direction, and I should like to quote from a letter in which she describes her method.

> These images work rather like a kaleidoscope, an instrument which contains bits and pieces by means of which structural patterns are realized. What is produced is a net of relationships. The images come to function both as systems of abstract relations and as objects of contemplation. These abstract relations are definable by the number and nature of the axes employed; for example, the fragments have to be all alike in various respects, such as size, shape, brightness or colouring, or to partake of a common quality, like having spots or stripes or all being smooth or all with wings or all ten foot high . . . It is a way of relating different but interwoven scales and dimensions.

Ruskin wrote of pictures: 'You must consider the whole as a prolonged musical composition.' Among all the arts music has traditionally been the medium for the purest expression of abstract structural relations. And 'rhyme', in the form of thematic variation, emerges as music's fundamental principle – the stock-in-trade of nearly every musical composer. The composer presents us with, say, a simple melody, repeats it a few times and then launches into a series of variations, playing it on a different instrument, with different emphasis or in a different key, until eventually he returns to the original. But repetition of the same theme, albeit with variations, becomes in the long run relatively dull. As in poetry – and in every other 'taxonomic' activity – *contrast* is needed to bring home the unity of the rhyming elements, and the composer typically introduces a contrasting theme with its own variations. Thus in a simple piece such as a Chopin nocturne two distinct themes, I and II, are arranged in the following way: I–I–II–I–II–I. To take the nocturne in E flat as an example, the first tune is repeated twice so that the main key and the main subject-matter may be well established in the memory of the hearer. Then comes the second tune, which is in the most nearly related key (so that the effect of the contrast is not lost because of too great dissimilarity). Then the two tunes alternate, while at each repetition small changes are introduced, in the form for instance of decorative arabesques in the right-hand part. In more complex pieces, such as Beethoven sonatas, the composer introduces a 'development' section in which the motifs of the first theme are picked up and rearranged until, just at the point where the hearer may be in danger of losing track of what is going on, order is restored by the 'recapitulation' of the first theme pure and simple.

'Sonata form' is to my mind a perfect example of an instructive and challenging exercise in classification. If I were an educational psychologist concerned with developing teaching machines for use in schools I would not, as the American behaviourists have done, base my machines on principles derived from experiments on how pigeons perform in Skinner-boxes, but instead would turn directly to the hallowed principles of musical design.

And that brings me close to the end of this chapter – but to an end, I should say, which is not altogether in tune with the beginning. I began by saying that man-made beauty is a lie. And so, in a sense, I believe it is. It was not for the sake of Beethoven sonatas that men evolved to take such a delight in classifying; to that extent, Beethoven merely capitalised on a human faculty which was developed for quite other reasons. But though it may be true that, at one level, we gain nothing of biological value from learning to classify the themes of a sonata, it does not necessarily follow that listening to a sonata has no function at all. For it can be argued that, at another level, through the experience of beauty in works of art we *learn to learn*. I implied as much just now when I suggested that psychologists might use music as a model for the design of teaching machines. If psychologists could learn from music how best to present evidence of the relations between things in a way that people will find easy to take in, it is equally possible that laymen might learn to do it for themselves. In that way the work of art would achieve a new importance, as a model of the way we should structure our experience wherever and whenever we need to acquire genuinely useful information.

Everyone has heard the argument that a training in Latin and Greek, however irrelevant to real life, is an excellent 'training for the mind'. How much better for that purpose might be a training in the appreciation of beauty, if, as I've argued, our love of beauty has a hundred million years of educational psychology behind it.

Let me finish, then, on a more positive note. Beauty may be an 'illusion'. But, for all that, Keats was not so wrong in his claim that 'Beauty is truth, truth beauty.' That may not be 'all ye know on Earth, and all ye need to know', but it is at least a good beginning.

10

TURNING THE LEFT CHEEK

with Christopher McManus

Van Gogh's picture *The Potato-Eaters* (Plate 4) shows a family of peasants seated at a meal. Imagine now that the picture were reversed, as in a mirror. Although every detail of the picture would change hands, it might be thought that *as a work of art* the mirror-image would be no less satisfying than the original. We know, however, that this was not Van Gogh's opinion. Having made a preliminary sketch for his picture, he made a lithograph. To do so he copied the sketch straight on to the block and the print he made turned out therefore to be a mirror image. His dissatisfaction with the print is recorded in a letter he wrote to his brother Theo: 'If I make a picture of the sketch, I shall make at the same time a new lithograph of it, and in such a way that the figures which, I'm sorry to say, are now turned the wrong way, come right again.'[1]

What is it about the mirror image of a picture that could lead to its being described as 'wrong'? On the part of the artist it might be simply that he is less familiar with the new version and thus likes it less. Yet it has been shown that people (artists in particular) can generally pick out the mirror image of a picture from the original when the picture is wholly unfamiliar, provided both versions are presented together. Left–right asymmetry is not, it seems, a neutral, 'accidental' feature of the composition.

We decided to examine asymmetry in painted portraits. Portraits are particularly suitable material for such a study: first, because there are plenty of them (produced to a rather standard pattern for rather standard ends); and, second, because the asymmetry in portraits is of a relatively simple kind which is obvious to the eye and easily classified. Portrait painters rarely paint their subjects full-face, but rather turn the head slightly to one side so that a more recognisable 'three-dimensional' image is produced. In so far as the painter's goal is merely to portray the physical likeness of the sitter, there would seem to be no reason for any consistent bias to turn the head to left or right. Thus we might expect that, in a sufficiently large sample, 50 per cent of the portraits would show more of the left cheek and 50 per cent more of the right. That was the 'null hypothesis' with which we started. We soon found good cause to reject it.

We examined 1,474 painted portraits, each showing a single person only, produced in western Europe from the fourteenth to the twentieth century. The sources were the National Portrait Gallery in London, the Fitzwilliam Museum in Cambridge, Roy Strong's definitive textbook on Elizabethan and Jacobean portraiture, and a miscellaneous collection of other books on art. Of this large sample of portraits, we found that 891 showed more of the left cheek and 583 more of the right. Such a ratio – 60 per cent to 40 per cent – would be expected to occur by chance less than once in 10,000 times.

The most immediate explanation of this bias (and the one that was favoured by every art historian to whom we talked) is that it is due to some mechanical factor related to right-handedness. Thus, it might be that a right-handed artist simply finds it somewhat easier to draw a profile to the left of the canvas. But we discounted this explanation when we analysed further the results of our survey.

We divided the portraits according to the sex of the subject and found that although 68 per cent of the women showed more of the left cheek, only 56 per cent of the men did so: both results are significantly different from chance – but, more importantly, the difference between the men and women is itself highly significant statistically. It is difficult to conceive of any purely mechanical explanation which can convincingly account for this sex difference.

Further grounds for discounting the importance of handedness are provided by the finding that the bias is much less marked in portraits showing the face in full profile than in those showing it in three-quarter profile; a handedness explanation would probably predict the reverse. Also, the bias is less marked in portraits showing only the head and shoulders than in those which show the rest of the subject's body.[2]

Some other explanation is called for. Yet the results, as they stood at that stage, did not – we were bound to admit – give solid support to any alternative explanation we could think of. There was one possibility, however, which seemed at least to hold some promise: an explanation in terms of left–right symbolism.

The gist of the idea was that the portrait painter uses 'left' and 'right' as signs to convey information about the sitter's character or status (perhaps without being consciously aware of doing so). Such sign-systems are known to operate in several types of primitive art. In the magical drawings (called 'ongons') of the Mongolian Buryat people, for instance, the social and spiritual status of the figures in the drawing is indicated by the figures' position up and down, and left and right. We thought that the professional portrait artist, constrained by his client to produce an accurate (and flattering) representation, might perhaps use the turn of the head to make a more personal statement about the sitter's status. We have since obtained further evidence which supports this theory.

We considered the work of one artist alone rather than pooling the data from many artists, hoping thereby to bring out the particular factors which influenced the individual artist making a decision as to how his subject should face. We chose the work of Rembrandt van Rijn.

Rembrandt is known to have painted well over 300 portraits, including 57 self-portraits (roughly two a year for the whole of his working life). Taking first the self-portraits, we found that nine showed more of the left cheek and 48 more of the right, as in Plate 2 – that is only 16 per cent showing the left cheek. However, the rest of the portraits revealed a very different pattern. We categorised the portraits according to the sex of the subject and the kinship relation to Rembrandt himself – kin being his mother, father, sister, brother, wife (Saskia), mistress (Hendrijke Stoffels) and son (Titus). The results of this analysis are shown in Figure 5. The single most important result to emerge, apart from the sex difference, is that portraits of non-kin are much more likely to show the left cheek (as in Plate 3) than portraits of kin, the difference being significant at the one in fifty level.

How do we interpret this remarkable finding? We suggest quite simply that Rembrandt structured his social world along the dimension 'socially like myself/socially unlike myself' and his portraits along the dimension 'showing the right cheek/showing the left cheek'. In Rembrandt's mind these two constructs were, in George Kelly's terminology, parallel and equivalent.

Thus whenever Rembrandt painted a portrait he gave some indication of the social distance between himself and the subject. He felt that the group of subjects most like himself were his male

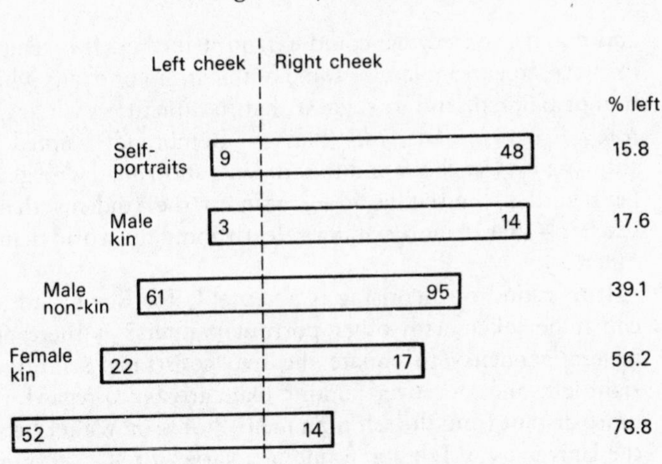

Figure 5. Analysis of 335 portraits by Rembrandt.

kin, and after them came the male non-kin. He considered women in general to be much less close to him, even to the extent that his female kin were more remote than the non-kin males. This appears to be true even of his wife, of whom 60 per cent of the 15 portraits show the left cheek.

Such attitudes are perhaps to be expected of a man of Rembrandt's position in the Holland of the seventeenth century. We might thus reconstruct for Rembrandt a kind of palaeo-psychology;

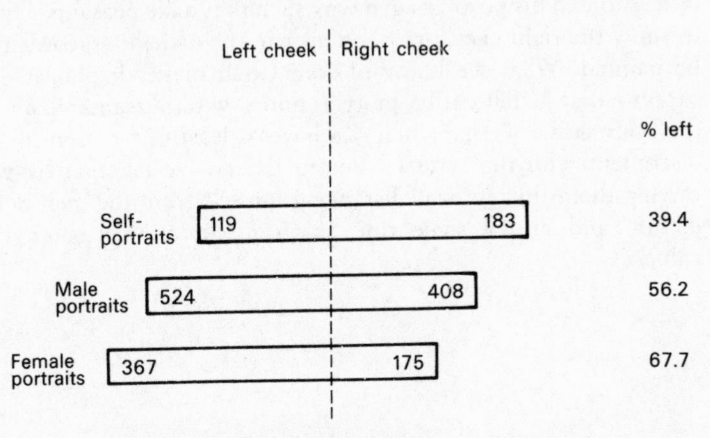

Figure 6. Analysis of 1,776 portraits by many painters.

and maybe, indeed, we could extend it further. It is tempting, for instance, to extrapolate to some of the group portraits which Rembrandt painted, and to suggest that in a picture such as *The Anatomy Lesson of Dr Tulp* (Plate 5) Rembrandt painted Dr Tulp showing his left cheek and the students showing their right because he regarded himself as more akin to the students than to the teacher – that is, more willing to learn about the world than to teach others.

If this kind of theorising is acceptable for Rembrandt, how far can it be taken with other portrait painters? Is there perhaps a general tendency to equate the two constructs self/non-self and right/left, and moreover (among male artists) to regard women as more distant from the self than men? Professor Walter Landauer of the University of London found in a survey of 302 self-portraits by different artists that 39 per cent showed the left cheek. When these data are put together with those from our earlier survey we get the pattern of results shown in Figure 6. The general similarity between the data for Rembrandt alone (Figure 5) and for many artists collectively (Figure 6) is so striking that we cannot help feeling that the analysis given for Rembrandt may have more universal validity.

Men/women and kin/non-kin are but two of the dimensions which could, in principle, be correlated with the dimension self/non-self. We should expect each individual artist to have his own notions of what kind of people were close and what remote from him. There is suggestive evidence that Van Gogh, for example, differentiated his portraits in a way such that male peasants tended to show the right cheek more often than the male bourgeoisie that he painted. What we know of Van Gogh makes it plausible to suppose that he felt closer, more at home, with peasants than with the bourgeoisie. Perhaps here we have at least one reason for his discontent with the reversed Potato Eaters. He felt that by portraying them thus he was distancing himself from the men in the picture and at the same time imposing upon them bourgeois values.[3]

11
BUTTERFLIES THAT STAMP

In 1974 Charles Sibley of Yale University was fined $3,000 for 'illegally importing bird parts taken abroad in violation of foreign wild life laws'. The 'bird parts' in question were eggs. Sibley's prosecution was the result of detective work carried out by Richard Porter, Investigations Officer of the Royal Society for the Protection of Birds and scourge of British egg-collectors. Porter himself expressed bewilderment at the activities of the men he hunted down. 'Why do they do it?' he asked. 'They camp out in appalling conditions. They carry out a desperately dangerous climb sometimes. Then they put the eggs in a cabinet with just themselves and a few friends to see them. They're just like kids.'[1]

'Everyone knows', as Pavlov noted, 'the dedication, the single-mindedness with which the collector achieves his purpose.' But while egg-collectors and stamp-collectors surprise us by their passion for acquiring worthless things, everyone knows, too, certain people's devotion to a form of collecting that is in some ways even more bizarre: I mean their devotion to the activity of 'spotting', to collecting nothing more than *observations*. When I was a boy I spent long hours 'loco-spotting', standing on a railway bridge to record the number of each train as it went by. And most children at one time or another have gone out armed with an 'Observer Book' or 'I-Spy' manual, with no grander purpose than merely to *catch sight of* a bird, toadstool, licence-plate, pub-sign or whatever. The craze may pass in childhood. But it may not. The 'aero-spotters', middle-aged as well as middle-brow, who throng the terminal roofs at Heathrow Airport continue to provide big business for the London shops which cater for their needs. 'The typical aero-spotter does nothing other than aero-spot,' the proprietor of a shop for spotters told the *Guardian* newspaper. 'He isn't married, or anything like that.'[2]

That a man should become so engrossed in a pursuit that he risks his life on a mountain, wastes his time on a railway bridge, or denies himself the possibility of marriage, needs explanation. Why does he – why do we – do it? There must presumably be something special about a Collection – a Collection, that is, with a big C.

What is a Collection? How does it differ from any mere aggregation of bizarre or precious objects? Collectors are catholic in their tastes: stamps, snuff-boxes, butterflies – almost anything will do. Yet we should note that no single collection contains all these things. The first principle of collecting is that the objects collected should fall within a strictly bounded category: a cigarette-card has no place in a stamp-collection, nor a snuff-box in a collection of thimbles, however rare and valuable in its own right each may be. The same goes for spotting: an aero-spotter will probably not so much as notice the hen-harrier hovering above the Hawker Harrier on the runway.

But there seems to be a second principle, in some ways contrary. For while it's true that all the members of a collection must be similar, the fact remains that none may be so similar as simply to duplicate another. Fifty identical stamps do not make a stamp-collection, and fifty sightings of the same railway engine bring no joy to the loco-spotter. Thus within the limits of his chosen category, the collector typically searches for and prizes *variation*. 'Look at the crested titmouse eggs,' said one of Professor Sibley's egg-collecting friends. 'This one belongs to the Scottish crested titmouse, this one to the Continental titmouse. Look at the *difference*.'

So a collection is distinguished from a mere accumulation by its structure. It must have unity, it must have variety. And the relationships between the elements become as important as the elements themselves: they must have what Gerard Manley Hopkins (see Chapter 9) called *rhyme*.

This much, I suspect, is obvious. But its obviousness should not blind us to its meaning. The recognition that collecting is an activity defined by clearly statable abstract rules allows us to dismiss at once the idea that it reflects simply some kind of acquisitive or hoarding tendency. There are, admittedly, collectors to whom the *owning* of a collection has become all-important, and they might seem to have little in common with the spotters who simply record their observations in their notebooks or hold them in their heads. But it would be wrong to pay too much attention to ownership as such. Although commercial exploitation has in recent years added a new dimension to collecting, the mercenary aspect should be seen for what it is, the prostitution of a more innocent and basic human drive. Collectors of objects – as much as collectors of observations – are usually (and primarily) in it not for the material satisfaction, but for the mental thrill it gives them. The pleasure they get from it is linked to the satisfaction human beings take in classifying incoming information: making comparisons, uncovering relationships, and imposing order on the world.

The biological reasons why human beings, along with other animals, should have developed a need – and a love – for classifying experience have been explored in more detail in Chapter 9. But what should we say about a biological trait which leaves people 'not married, or anything like that' – whose effect, *in extremis*, is to reduce a man's biological fitness to zero? Only, perhaps, that where nature has sown the wind, modern human beings can and will reap the whirlwind. The 'environment of adaptiveness' of human beings, the environment where they developed the instinct to be collectors of rhyming observations, is not the environment they live in now. In civilised society man-made artefacts such as postage stamps and aeroplanes, which fall so neatly into quasi-natural categories, will capture and pervert the natural human tendency to look for order. The aero-spotter or the stamp-collector may indeed be as much the dupe of an artificial 'super-stimulus' as the black-headed gull who sits on a gigantic china egg in preference to her own.

12

THE COLOUR CURRENCY OF NATURE

Man as a species has little reason to boast about his sensory capacities. A dog's sense of smell, a bat's hearing, a hawk's visual acuity, are all superior to man's own. But in one respect he may justifiably be vain: in his ability to see colours he is a match for any other animal. In this respect he has surprisingly few rivals. Among mammals only his nearest relatives, the monkeys and apes, share his ability – all other mammals are nearly or completely colour-blind. In the rest of the animal kingdom colour vision occurs only in some fishes, reptiles, insects and birds.

No one can doubt man's good fortune. The world seen in monochrome would be altogether a drearier, less attractive place to live in. But nature did not grant colour vision to man and other animals simply to indulge their aesthetic sensibilities. The ability to see colour can only have evolved because it contributes to biological survival.

The question of how colour vision has evolved is – or should be – an important issue for psychologists (and for designers). If we were to understand how the seeing of natural colour has in the distant past contributed to man's life we might be better placed to appreciate what colour in 'artificial' situations means to him. Yet it is not in fact an issue which has been much explored. Indeed, few psychologists, for all their obsession with the physiological mechanism of colour vision, have asked what to an evolutionary biologist must seem the obvious question: Where – and *why* – does colour occur in nature?

It may seem odd to tack 'Why?' on to the question 'Where?' But the question *why* is crucial, for the evolution of colour vision is intimately linked to the evolution of colour on the surface of the earth. It may go without saying that, in a world without colour, animals would have no use for colour vision; but it does need saying that in a world without animals that possessed colour vision there would be very little colour. The variegated colours which characterise the earth's surface (and make the earth perhaps the

Plate 1. From *Tropisms* (1970), a series of twelve collage-paintings by Suzi Gablik.

Plate 2. Self-portrait by Rembrandt, 1653.

Plate 3. Portrait by Rembrandt of Margaretha van Bilderbeecq, 1633.

Plate 4. *The Potato-Eaters* by Van Gogh.

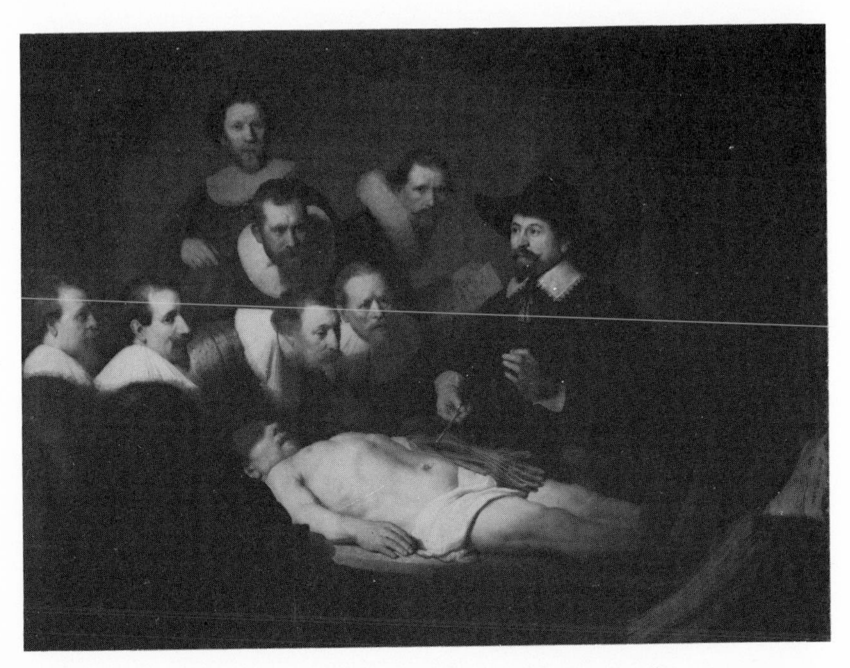

Plate 5. *The Anatomy Lesson of Dr Tulp* by Rembrandt.

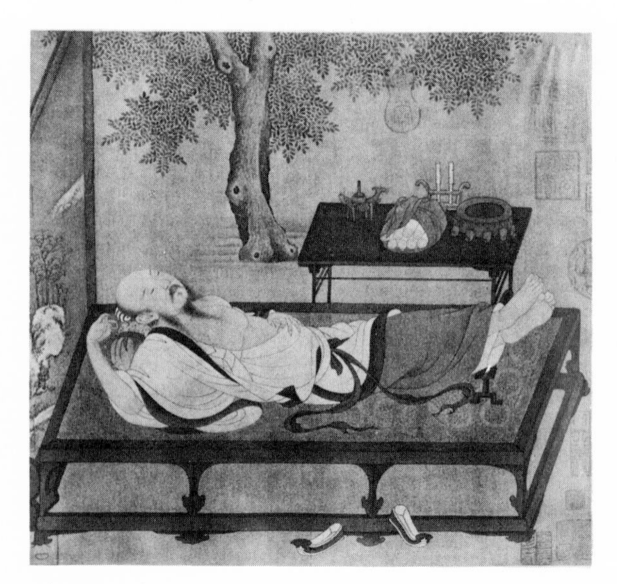

Plate 6. *Midsummer Rest under a Locust Tree*, northern Sung.

Plate 7. *The Resurrection* by Ugolino di Nerio.

most colourful planet of the universe) are in the main *organic* colours, carried by the tissues of plants and animals – and most of these life-born colours have been designed in the course of evolution to be *seen*.

There are of course exceptions. Before life evolved, the drab landscape of the earth may have been relieved occasionally by, say, a volcanic fire, a rainbow, a sunset, perhaps some tinted crystals on the ground. And before colour vision evolved some living tissues were already 'fortuitously' coloured – blood was red, foliage green, though the redness of haemoglobin and the greenness of chlorophyll are wholly incidental to their biochemical roles. But the most striking colours of nature, those of flowers and fruits, the plumage of birds, the gaudy fishes of a coral reef, are all 'deliberate' evolutionary creations which have been selected to act as visual *signals* carrying messages to those who have the eyes to see them. The pigments which impart visible colour to the petals of a dandelion or a robin's breast are there for no other purpose.

We may presume that colour vision has not evolved to reveal the rare colours of inorganic nature, since rainbows and sunsets have no importance for survival. Nor is it likely to have evolved so that it is possible to see simply the greenness of grass or the redness of raw flesh, since those animals which feed chiefly on grass or on flesh are colour-blind. It has surely evolved alongside signal coloration to enable animals to detect and interpret nature's colour-coded messages.

The messages conveyed by signal coloration are of many kinds. Sometimes the message is simple: 'Come here' addressed to an ally (the colour of a flower serving to attract a pollinating insect, the colour of a fruit to attract a seed-dispersing bird), or 'Keep away' addressed to an enemy (the colour of a stinging insect or a poisonous toadstool serving to deter a potential predator). Sometimes the message is more complex, as when colour is used for communication in a social context, in courtship or aggressive encounters (a peacock displaying his fan, a monkey flashing his coloured genitalia). Whatever the level of the message, signal colours commonly have three functions: they catch attention, they transmit information, and they directly affect the emotions of the viewer – an orange arouses appetite in a monkey, a yellow wasp fear in a fly-catcher, the red lips of a young woman passion in a man.

Primates came on the scene relatively late in evolutionary history, and the surface of the earth must already have been given much of its colour through the interaction of plants, insects, reptiles and birds. The early tree-dwelling primates moved in on an ecological niche previously occupied by birds: they picked the same fruits, caught the same insects, and they were in danger of being harmed by the same stings and the same poisons. To compete effectively with birds, primates needed to evolve colour vision of the same order. It is for that reason, I suspect, that the red-green-blue system of colour vision possessed by most primates (including man) is in fact so similar to that, say, of a pigeon (although, as it happens, the selectivity of the three types of colour receptor is achieved by quite different physiological mechanisms in primates and birds). Once primates had joined the colour-vision club, however, they too must have played their part in the progressive evolution of natural colour, influencing through selection the colours both of themselves and of other plants and animals.

Then, not far back in history, the emergence of man marked a turning-point in the use of colour. For men hit on a new and unique skill – the ability fo *apply* colour in places where it did not *grow*. Most probably they first used artificial colour to adorn their own bodies, painting their skins, investing themselves with jewels and feathers, dressing in coloured clothes. But in time they went further and began to apply colour to objects around them, especially to things which they themselves had made, until the use of colour became eventually almost a trademark of the human species.

In the early stages men probably continued the natural tradition of using colour primarily for its signal function, to indicate maybe status or value. And to some extent this tradition has continued to the present day, as testified, for instance, in the use we make of colour in ceremonial dress, traffic-signals, political emblems, or the rosettes awarded to horses at a show. But at the same time the advent of modern technology has brought with it a debasement of the colour currency. Today almost every object that rolls off the factory production line, from motor cars to pencils, is given a distinctive colour – and for the most part these colours are *meaningless*. As I look around the room I'm working in, man-made colour shouts back at me from every surface: books, cushions, a rug on the floor, a coffee-cup, a box of staples – bright blues, reds, yellows, greens. There is as much colour here as in any tropical forest. Yet while almost every colour in the forest would be meaningful, here in my study almost nothing is. Colour anarchy has taken over.

The indiscriminate use of colour has no doubt dulled man's biological response to it. From the first moment that a baby is given a string of multi-coloured – but otherwise identical – beads to play with he is unwittingly being taught to *ignore* colour as a signal. Yet I do not believe that our long involvement with colour as a signal in the course of evolution can be quite forgotten. Though modern man's use of colour may frequently be arbitrary, his response to it continues to show traces of his evolutionary heritage. So men persist in seeking meaning from colour even where no meaning is intended: they find colour attention-catching, they expect colour to carry information and to some extent at least they tend to be emotionally aroused.

The most striking illustration of man's biological inheritance is the significance which is attached to the colour *red*. I was first alerted to the peculiar psychological importance of red by some experiments not on men but on rhesus monkeys. For some years I studied the visual preferences of monkeys, using the apparatus shown in Figure 7. The monkey sits in a dark testing-chamber with a screen at one end on to which one of two alternative slides can be projected. The monkey controls the presentation of the slides by pressing a button, each press producing one or the other slide in strict alternation: thus when he likes what he sees he must hold the button down, but when he wants a change he must release and press again. I examined 'colour preference' in this situation by letting the monkeys choose between two plain fields of coloured light. All the monkeys tested showed strong and consistent preferences. When given a choice between, for instance, red and blue,

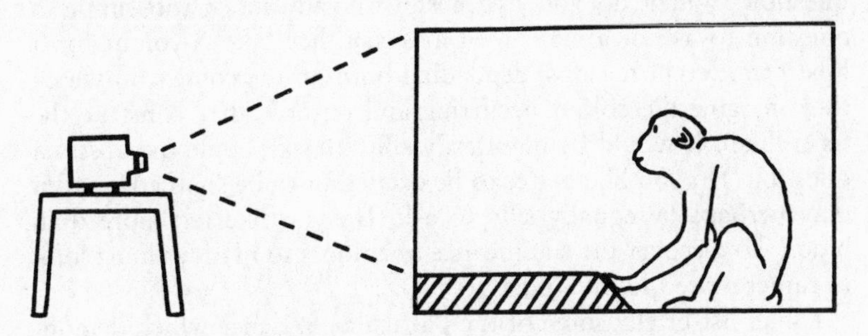

Figure 7. Apparatus for testing monkeys' visual preferences.

they tended to spend three or four times as long with the blue as with the red. Overall, the order of preference was blue, green, yellow, orange, red. When each of the colours was separately paired with a 'neutral' white field, red and orange stood out as strongly aversive, blue and green as mildly attractive.[1] Direct observation of the monkeys in the testing situation indicated that they were considerably upset by the red light. When I deliberately added to their stress by playing loud and unpleasant background noise throughout the test, the aversion to red light became even more extreme. Further experiments showed that they were reacting to the red light exactly as if it was inducing fear.[2]

This aversion to red light is not unique to rhesus monkeys. The same thing has been found with baboons and also, more surprisingly, with pigeons. But what about men? Experiments on colour preference in men have given results which appear at first sight to be at odds with those in other primates. When men are asked to rank colours according to how much they 'like' them red often comes high if not top of the list, although there is a wide variation between individuals, depending among other things on personality, age, sex and culture. I am inclined to give little weight to such findings for two reasons. First, the choice of a 'favourite' colour may be heavily biased by changes in fashion; indeed, when T. Porter tested people from social backgrounds where fashion probably has relatively little influence – African children on the one hand, the residents of an Oxford old people's home on the other – he found that both groups ranked colours in much the same way as did my monkeys, consistently preferring the blue end of the spectrum to the red.[3] Second, and more important, there is a methodological problem with most of the preference experiments. The question 'Which do you *like* best?' is really much too simple a question to ask of a man: men may say they 'like' a colour for a host of different reasons, depending both on the context in which they imagine the colour occurring and on how they construe the term 'like'. It would be manifestly silly to ask people the abstract question 'Do you like better to be excited or to be soothed?' and it may perhaps be equally silly to ask 'Do you like red more than blue?' To discover the significance of colours to man we must look to rather more specific studies.

I shall list briefly some of the particular evidence which demonstrates how, in a variety of contexts, red seems to have a very special significance for man. (i) Large fields of red light induce physiological symptoms of emotional arousal – changes in heart-rate, skin resistance and the electrical activity of the brain.[4] (ii) In patients suffering from certain pathological disorders, for instance cerebellar palsy, these physiological effects become exaggerated – in cerebellar patients red light may cause intolerable distress, exacerbating their disorders of posture and movement, lowering pain thresholds and causing a general disruption of thought and skilled behaviour.[5] (iii) When the affective value of colours is measured by a technique, the 'semantic differential', which is far subtler than a simple preference test, men rate red as a 'heavy', 'powerful', 'active', 'hot' colour.[6] (iv) When the 'apparent weight' of colours is measured directly by asking men to find the balance-point between two discs of colour, red is consistently judged to be the heaviest.[7] (v) In the evolution of languages, red is without exception the first colour word to enter the vocabulary – in a study of 96 languages B. Berlin and P. Kay found 30 in which the only colour word (apart from black and white) was red.[8] (vi) In the development of a child's language red again usually comes first, and when adults are asked simply to reel off colour words as fast as they can they show a very strong tendency to start with red.[9] (vii) When colour vision is impaired by central brain lesions, red vision is most resistant to loss and quickest to recover.[10]

These disparate facts all point the same way, to the conclusion that man as a species finds red both a uniquely impressive colour and at times a uniquely disturbing one. Why should it be so? What special place does the colour red have in nature's scheme of colour signals?

The explanation of red's psychological impact must surely be that red is by far the most common colour signal in nature. There are two good reasons why red should be chosen to send signals. First, by virtue of the contrast it provides, red stands out peculiarly well against a background of green foliage or blue sky. Second, red happens to be the colour most readily available to animals for colouring their bodies because, by pure chance, it is the colour of blood. So an animal can create an effective signal simply by bringing to the surface of its body the pigment already flowing through its arteries: witness the cock's comb, the red bottom of a monkey in heat, the blush of a woman's cheek.

The reason why red should in certain situations be so disturbing is more obscure. If red was always used as a warning signal there would be no problem. But it is not: it is used as often to attract as to repel. My guess is that its potential to disturb lies in this very *ambiguity* as a signal colour. Red toadstools, red ladybirds, red poppies, are dangerous to eat, but red tomatoes, red strawberries, red apples, are good. The open red mouth of an aggressive monkey is threatening, but the red bottom of a sexually receptive female is appealing. The flushed cheeks of a man or woman may indicate anger, but they may equally indicate pleasure. Thus the colour red, of itself, can do no more than alert the viewer, preparing him to receive a potentially important message; the content of the message can be interpreted only when the *context* of the redness is defined. When red occurs in an unfamiliar context it becomes therefore a highly risky colour. The viewer is thrown into conflict as to what to do. All his instincts tell him to do *something*, but he has no means of knowing what that something ought to be. No wonder that my monkeys, confronted by a bright red screen, became tense and panicky: the screen shouts at them 'this is important', but without a framework for interpretation they are unable to assess what the import is.[11] And no wonder that human subjects in the artificial, contextless situation of a psychological laboratory may react in a similar way. A West African tribe, the Ndembu, state the dilemma explicitly: 'red acts both for good and evil'.[12] It all depends.

I have tried to show how an evolutionary approach can help throw light on man's response to colour. Whether this approach can be helpful to the practice of design remains an open question. In many areas of our lives we already overrule and nullify man's natural tendencies. But I believe we should try to be 'conservationists' as much on behalf of ourselves as we are learning to be on behalf of other species, and that we should try where possible to make our style of life conform to the style to which man is biologically adapted. Designers, who are now more than anyone responsible for colouring our world, have a choice before them. They can continue to devalue colour by using it in an arbitrary, non-natural way, or they can recognise and build on man's biological predisposition to treat colour as a signal. If they choose the latter, bolder, course they might do well to study how colour is used in nature. Nature has, after all, been in the business of design for over a hundred million years.

13
CONTRAST ILLUSIONS IN PERSPECTIVE

In the house of the early Cubist painters in Paris lived a certain Maurice Princet, a somewhat derelict figure known to the group as Le Mathématicien. Princet is said to have addressed the following question to Picasso and Braque: 'You represent by means of a trapezoid a table as you see it transformed by perspective, but what would happen if the fancy struck you to express the table as a type? It would be necessary for you to place it on the plane of the canvas, and return from the trapezoid to the true rectangle.'[1]

But had he tried this formula himself, Princet, the mathematician, might have been surprised at the result. Figure 8 shows a table, not 'transformed by perspective' so that its top is a trapezoid, but rather 'expressed as a type' so that its top is a perfect parallelogram. Curiously, it does not look like a parallelogram, but like a trapezoid. We are subject to a visual illusion, an illusion new to science.

The conditions responsible are easily discovered. Removing the legs from the table makes the illusion almost disappear; placing on it a vase of flowers makes it become more obvious still. The error we make in judging the shape of the parallelogram depends on our

Figure 8. The table-top illusion. See text for explanation.

seeing it as a table-top, with the implication that it represents a receding horizontal surface. It so happens that whenever two equal lines are convincingly depicted in such a way that one appears more distant than the other, the further line looks longer than the nearer. But why should apparent depth have such an influence on perception of size?

When people make estimates of the value of sensory stimuli it is generally true that their estimate depends not only on the actual value of the stimulus but also on what they expect its value to be. Deviations from the expected value are exaggerated, so that if the value of a stimulus is less than expected it is underestimated, and if it is greater it is overestimated – the greater the discrepancy, the greater the error.

In this formulation, expected value has rather a special meaning. Other things being equal, the 'expectation' is conditioned simply by the prevailing sensory context, and the expected value approximates to the average value of similar stimuli in the local environment. In the presence of large stimuli the expected value is high and in the presence of small stimuli it is low. Hence a particular stimulus is judged to be smaller in the former context than the latter.

These so called 'contrast effects' occur in all sensory modalities. A light of a particular brightness, say a torch-bulb, is judged to be dimmer in a brightly lit room than in a dark one; an interval of a particular duration, say a second, is judged to be longer in the fast movement of a symphony than in the slow one; a fruit of a particular sweetness, say an orange, is judged to be sourer when eaten with a sweet drink than with a sour one. Less familiar examples occur in the perception of visual form. Figures 9 and 10 show the influence of contrast on the size of a circle and the orientation of a line: a circle is judged to be smaller when surrounded by large circles than by small; a line is judged to slope away from another line which crosses it.

The effect of a particular sensory context may persist to some extent when the context is removed, so that the subjective estimate of a particular stimulus is influenced not only by current but by previous exposure to stimuli of differing value. After wearing red glasses a piece of white paper seems to be tinged with green; after speeding along a motorway we judge a limit of 30 miles per hour to be dreadfully slow; after stepping off a heaving ship we feel that the ground under our feet is going up and down. Such after-effects, too,

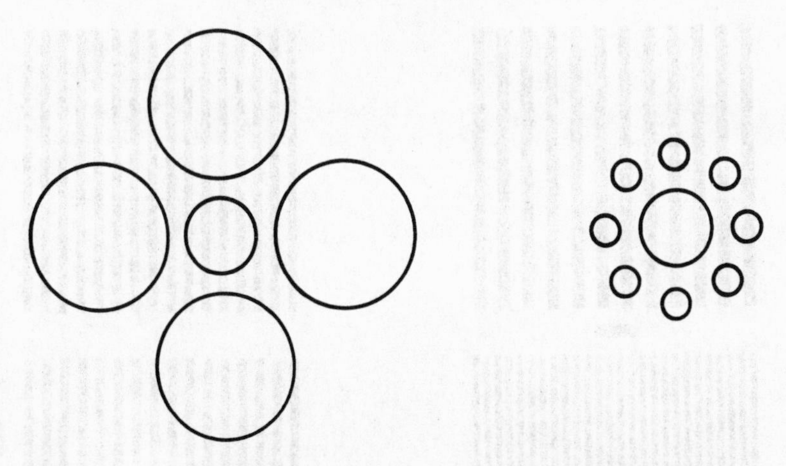

Figure 9. A contrast illusion for size. The two central circles are equal in diameter.

occur for visual form. The effect on the perception of size can be shown with Figure 11. Here, prolonged inspection of the left-hand

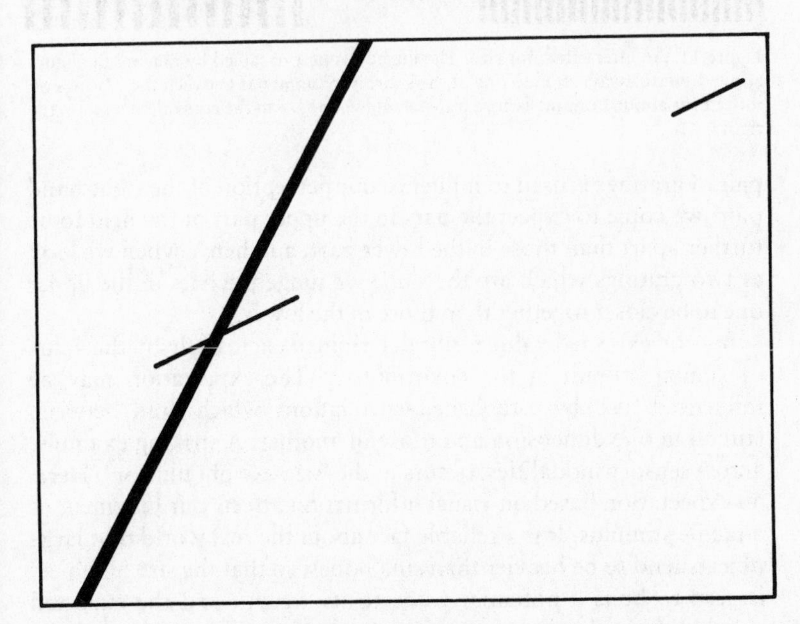

Figure 10. An orientation illusion. The short line at the top on the right is a continuation of the shorter of the lines that cross each other on the left.

Figure 11. An after-effect for size. The illusion is best obtained by placing the figure about 1 metre away, and looking at the short horizontal bar between the gratings on the left for about 1 minute before transferring your gaze to the equivalent spot on the right.[2]

pair of gratings is used to influence our perception of the right-hand pair: we come to expect the bars in the upper part of the field to be further apart than those in the lower part, and hence when we look at two gratings which are the same we judge the bars in the upper one to be closer together than those in the lower.

But the expected value is not determined exclusively by the value of similar stimuli in the environment. The expectation may be influenced also by established correlations which hold between stimuli in one dimension and those in another. A striking example, across sensory modalities, occurs in the 'size–weight illusion'. Here, an expectation based on visual information affects our judgment of a tactile stimulus. It is a reliable fact about the real world that large objects tend to be heavier than small ones, so that the size an object is seen to be is a potential guide to its weight, and the expected weight of the larger of two objects is greater than that of the smaller. Hence if we are asked to lift two objects of different size but of the same actual weight, we judge the larger to weigh less than the smaller – we judge the pound of feathers to be lighter than the pound of lead.

The 'depth–size illusion' of Figure 8 is, to my mind, a phenomenon of this latter kind, where information from one stimulus dimension conditions our expectation about the value of stimuli in another. Seeing the drawing *as* a table, and seeing the upper edge as *more distant* from us than the lower, we expect the upper edge to be *foreshortened* by perspective. But it is not. And the discrepancy between what we expect and what we observe leads us to exaggerate the 'error' and see the upper edge as if it were actually drawn longer on the page.

By this account, the depth–size illusion, the size–weight illusion and the illusions shown in Figures 9–11 are essentially similar phenomena, originating in the contrast between the actual and the expected values.

At a physiological level it is possible that these illusions have a common mechanism. The evidence of electrophysiology suggests that, for many stimulus dimensions, there are nerve units which respond preferentially to stimuli of certain values: any particular stimulus excites a range of such units, to a degree depending on how close their preferred value is to the stimulus's actual value, and the subjective estimate is based on an average of the overall response. The simplest explanation of the illusions is that those units which respond preferentially to the 'expected value' are selectively depressed. In the case of the conventional contrast illusions, where the expected value is determined simply by the value of other similar stimuli in the environment, this selective depression would result from adaptation and mutual inhibition among those units excited by such similar stimuli. In the case of the illusions where the expected value is determined by stimuli in another stimulus dimension, the depression would have to involve some kind of cross-dimensional inhibitory interaction.

Whatever fascination these illusions have for scientists concerned with the mechanisms of perception, their importance in practice might seem to be small. To graphic artists, however, the depth–size illusion may cause real trouble. In the painting called *Midsummer Rest under a Locust Tree* from the Peking Palace Museum (Plate 6) the tenth-century Sung artist has drawn both the couch on which the philosopher is lying and the table beyond as almost perfect parallelograms, with the result that the backs in fact look longer than the fronts. Again, in the painting of the Resurrection from the National Gallery in London (Plate 7) the Italian Primitive, Ugolino, has done the same with Christ's sarcophagus, and the illusion is present once more. Rather than using the laws of perspective, these artists may well have implicitly followed Princet's suggestion and drawn their material as true to a type instead of true to what they saw: the two sides of a couch, the two sides of a sarcophagus, are equal in reality – and they have been made equal on the page. We may guess that they were puzzled by the distortion which they found, but it would have been in keeping with medieval practice to prefer a formal rule to the mundane evidence of the senses.

The same cannot be said, however, of the Mongolian artist who drew the temple in Figure 12 to illustrate an eighteenth-century

Figure 12. Illustration from an eighteenth-century Mongolian text.

Buddhist text. He has drawn the far edge of the temple, which must have been meant to be square, not the same length as, but actually considerably longer than, the near edge. This way of doing things has seemingly no obvious rationale. But we can speculate about its history. What if this artist had innocently modelled his style on examples from the Sung school? He would have been tricked by his eyes into thinking that the Sung painter represented the more distant of two equal lengths as longer than the nearer. What was good enough for the Chinese would have been good enough for him, and in attempting to emulate them he has in fact given concrete expression to what was, unknown to him, the depth–size illusion.

OTHER PEOPLE'S BOOKS

14

AN ECOLOGY OF ECSTASY

The Spiritual Nature of Man
by Alister Hardy

Suddenly amid the sadness, spiritual darkness and depression, his brain seemed to catch fire, and with an extraordinary momentum his vital forces were strained to the utmost all at once. His sensation of being alive and his awareness increased tenfold . . . His mind and heart were flooded by a dazzling light. All his agitation, all his doubts and worries, seemed composed in a twinkling, culminating in a great calm, full of serene and harmonious joy and hope, full of understanding and the knowledge of the final cause.

Case 3001, Prince Myshkin, Male, Age 27.
Classification of experience: 1(b)(d); 7(a)(b)(f)(g)(i); 8(e); 9(a); 11(n); 12(a).

When Alister Hardy began his career in zoology at Oxford, the Revd W. A. Spooner was presiding over New College. Whatever the proclivities of his fellow students, it surely cannot have been Hardy of whom the story is told that he upset the Reverend Doctor by 'hissing all his mystery lectures, and tasting a whole worm'. I should be surprised if Hardy has ever hissed at anything, let alone mysteries; and I suspect that, if offered a whole worm, his natural tendency would always have been – after duly taking notes – to put it back in the earth where it belonged.

No one could have been better placed than this gracious, boyish, uncompromising scientist to found the Religious Experience Research Unit at Manchester College.

The possibility [he writes] of investigating man's transcendental experiences and of building up a body of knowledge about them from firsthand accounts has been a life-long interest which I have always regarded as part of my biological outlook. I began to collect material towards such a study more than fifty years ago: in September 1925 to be exact. I always regarded the planning of my research as an exercise in human ecology, for, to me, one of the greatest contributions biology could make to mankind would be to work out an ecological outlook which took into account not only man's economic and nutritional needs but also his emotional and spiritual behaviour.

To this end, beginning in 1969, Hardy launched a major survey of religious experience in contemporary Britain, soliciting through the Press and by means of questionnaires personal accounts from 'all those who feel that they have been conscious of, and perhaps influenced by, some Power, whether they call it God or not, which may either appear to be beyond their individual selves or partly, or even entirely, within their being'. The accounts would be treated in strict confidence, noted, classified and returned. No harassment – no hissing, no dissection . . . And the accounts poured in. At the end of eight years he had received over 3,000 firsthand descriptions of religious or quasi-religious experiences. A later random sample of the population by National Opinion Polls revealed that more than one in three people own to having been aware of or influenced by some such presence or power.

The Spiritual Nature of Man provides a summary of the Unit's findings, based on the first 3,000 cases. As Hardy saw it, his first job was to devise a system for classifying and labelling the different kinds of experience. He has devised an elaborate typology which allows any particular experience to be described under a variety of different headings: its sensory quality (visions, voices, out-of-the-body experiences . . .), its affective quality (sense of security, awe, ecstasy, sense of timelessness . . .), its antecedent cause or setting (prayer, music, natural beauty, despair, childbirth . . .), and so on. The greater part of the book is devoted to the explanation and illustration of how 'typical experiences' may thus be classified. The firsthand accounts, many of which Hardy quotes at length, are if nothing else remarkably frank, interesting, and very often moving; and the reader is likely to be astonished, as was Hardy himself, not only by the strangeness and richness of the experiences but by the wealth of literary expression which he has mined. The book contains a certain amount of statistical information relating to the relative frequencies of different kinds of experience (much of which appears to be mathematically invalid). And it concludes with an essay entitled 'What *is* spirituality?', in which Hardy signally fails to answer his own question, but makes it clear that whatever spirituality *is* he is in favour of it.

No hissing, no dissection, no attempt, in Gabriel Marcel's words (which Hardy quotes approvingly), to 'reduce mysteries to problems'. But *why not*? Why should we *not* ask, first, how much of all this stuff is genuine, and second, what it all means? If man does, perhaps, have a spiritual nature, he also for sure has a sceptical enquiring nature; and it is not obvious that we should agree to suppress the latter in order to do justice to the former. Hardy describes himself as a Darwinian. A Sunday Darwinian, maybe. From Monday to Saturday Darwin never left off making problems out of mysteries.

Hardy's defence against any deeper or wider enquiry into the significance of the material he has collected would be that at this, the first stage in developing a 'natural history of religious experience', the scientist's role must remain that of a detached observer. Collect the facts, classify them, and think about them later. Collect the facts, of course: provided they *are* facts, not fantasies. The facts that Hardy was after were presumably the facts about what people genuinely experienced at the time when the crisis, revelation or whatever came upon them. But what he actually collected were people's reports of what they remembered of their experiences. And that is a very different thing, especially when the informants were recalling events which occurred as much as half a century ago and when their modal age was between 60 and 69 years (not 50 and 59 years, as Hardy himself states, having miscalculated the numbers in his Table II). Memory plays tricks on people, particularly people who – on the evidence of their own accounts – are many of them of a hysterical disposition. What is more, people play tricks on scientists, especially when they are given such a golden opportunity either for self-glorification or for the glorification of their God. Anonymity, in this context, is no guarantee against a tendency on the part of informants to swank about their supposed religious experiences: God presumably knows the identity of each informant even if we, the readers of Alister Hardy's book, do not. Too many of the accounts read suspiciously like acts of worship in themselves.

Self-selection by informants is, as all sociologists know (and Hardy himself admits, only to dismiss it), a risky way of collecting information which is either honest or representative. As a basis for statistical analysis it is worthless. And yet, whatever reservations we may have about Hardy's method, we cannot dismiss his findings out of hand. Amid the more pompous and pious rubbish, there is unquestionably a good deal of material which does have the ring of truth to it. My real quarrel is with Hardy's failure to subject his strange material to any sort of constructive criticism. Having described and classified his 3,000 accounts he does little more than leave it at that, with a 'God bless' and an 'Isn't the world wonderful?'

Reverence for nature's works, although an admirable quality, is ultimately unenlightening. Would that Hardy showed a trace of irritation, that he tried – if not to uproot religious experience – at least to scratch around a little. But he does not. And the sceptical, problem-oriented reader is left to do it for himself.

When the gloss is removed, the most remarkable feature of these accounts of religious experience is their resemblance to pathological phenomena. As the book proceeds with its accounts of illuminations, guiding voices, senses of presence, depersonalisations, anyone with even a passing knowledge of medical case-histories is bound to recognise the tell-tale signs of epilepsy, migraine, schizophrenia, parietal-lobe brain-damage and so on. Indeed, anyone familiar with certain classic works of literary fiction is bound to do the same. I have quoted at the beginning of this review a passage from Dostoevsky's *Idiot*, where the author describes Prince Myshkin's feelings just before the onset of an epileptic fit (a passage for which there is an almost exact parallel in Dostoevsky's description of his own experience of epilepsy). For comparison let me quote one of Hardy's cases of religious experience:

> The phenomenon . . . is generally prefaced by a general feeling of gladness to be alive. I am never aware of how long this feeling persists but after a period I am conscious of an awakening of my senses. Everything becomes suddenly more clearly defined, sights, sounds and smells take on a whole new meaning. I become aware of the goodness of everything. Then, as though a light were switched off, everything becomes still, and I actually feel as though I were part of the scene around me.

Here is another case of Hardy's, for which the obvious parallel is the 'aura' of classical migraine:

> The experience lasted, I should say, about thirty seconds and seemed to come out of the sky in which were resounding harmonies. The thought: 'That is the music of the spheres' was immediately followed by a glimpse of luminous bodies – meteors or stars – circulating in predestined courses emitting both light and music. I stood still on the tow-path and wondered if I was going to fall down.

Here a case similar to what neurologists recognise as *kalopsia* (a sense of everything being beautiful and comforting), associated with lesions in the right parietal cortex of the brain:

> For a few seconds only, the whole compartment [of the train] was filled with light . . . I felt caught up into some tremendous sense of being within a loving, triumphant and shining purpose. I never felt more humble. I never felt more exalted. A most curious, but overwhelming sense possessed me and filled me with ecstasy. I felt that all was well for mankind. All men were shining and glorious beings who in the end would enter incredible joy. Beauty, music, joy, love immeasurable and a glory unspeakable, all this they would inherit.

Here is a case of out-of-the-body experience, in all respects like the *Doppelgänger* experience which is relatively common in epileptics, in patients suffering acute delirium or in those with damage to the parieto-occipital cortex:

> Later, when in my early twenties, as a lay preacher, I was taking a service in a tiny village chapel, I had another experience that remains both unusual and unique. Quite without any unusual context as I carried on the worship I ceased to be aware of my taking any active part. My only experience was that of sitting *behind* myself in the pulpit whilst I wondered how it came about that I was watching myself conducting the service.

I do not mean to imply that Hardy's informants were themselves seriously sick people. But it does seem likely that many of them at the time of their 'religious experience' were on the edge of sickness. Given, however, that the symptoms did not prove incapacitating, given that there was no one to *tell* them they were sick, the interpretation they put on their experience was religious. Better by far – within the context of our culture – to be inspired than to be insane.

In the eleventh century Hildegard of Bingen experienced a series of migraine-like attacks, associated with dramatic visual auras. One such vision she described as follows: 'I saw a great star most splendid and beautiful, and with it an exceeding multitude of falling stars which with the star followed southwards . . . And suddenly they were all annihilated, being turned into black coals . . . and cast into the abyss so that I could see them no more.' As Oliver Sacks writes in his book on *Migraine*, 'our literal interpretation would be that she experienced a shower of phosphenes in transit across the visual field, their passage being succeeded by a negative scotoma'.[1] But for Hildegard: 'The visions which I saw I beheld neither in sleep, nor in dreams, nor in madness, nor with my carnal eyes, nor with the ears of the flesh, nor in hidden places; but wakeful, alert, and with the eyes of the spirit and the inward ears, I perceived them in open view and according to the will of God.' And for Hildegard these visions were instrumental in turning her towards a life of holiness and mysticism. Only in recent years have experimental psychologists, drawing on the work of Schachter, begun to understand how physiological states of the body are given meaning and significance in people's minds by the context in which they arise and are interpreted: thus the very same bodily symptoms may be perceived by the person as fear, sexual love, anger, religious ecstasy, depending on the social and cultural setting.

Hardy must know all this. He must know of the parallels between his cases of 'religious experience' and cases of overt illness. He must know of the work of Schachter on the perception of emotions. He has been in the game too long, surrounded at Oxford by too many well-informed colleagues, for him not to know. Why, then, does he keep silent? Why has he written such an intellectually shallow book?

The answer, I suspect, lies in Hardy's misplaced anxiety not to hiss, not to dissect, never to undermine the illusion – even if it is an illusion – of divine intercession. Hardy knows from his own experience, and has had it confirmed often enough by the testimony of his informants, that those people who do 'feel that they have been conscious of, and perhaps influenced by, some Power, whether they call it God or not, which may . . . appear to be beyond their individual selves' are by that token happier, more confident, more generous beings. And, whom God – or cerebral physiology – has joined, let no sceptic put asunder.

But the danger is unreal. To turn a mystery into a problem does not reduce it, let alone destroy its value. On this point Dostoevsky's Idiot, having had the first, may perhaps have the last word:

> He arrived at last at the paradoxical conclusion: 'What if it is a disease? . . . What does it matter that it is an abnormal tension, if the result, if the moment of sensation, remembered and analysed in a state of health, turns out to be harmony and beauty brought to their highest point of perfection, and gives a feeling, undivined and undreamt of till then, of completeness,

proportion, reconciliation, and an ecstatic and prayerful fusion in the highest synthesis of life?' . . . If in that second – that is to say, at the last conscious moment before the fit – he had time to say to himself, consciously and clearly, 'Yes, I could give my whole life for this moment,' then this moment by itself was, of course, worth the whole of life.

15

STRAW GHOSTS

This House is Haunted: An Investigation of the Enfield Poltergeist
by Guy Lyon Playfair

Science and the Supernatural
by John Taylor

When I was an undergraduate at Cambridge I used sometimes to have tea with the philosopher C. D. Broad, and we would talk about ghosts. Professor Broad lived in Newton's rooms in Trinity, and his favourite spot for talking and for tea was an armchair placed beside the window where in 1665 Newton caught a beam of sunlight in the prism he had bought at Stourbridge Fair and spread it like a rainbow on the floor. Seated in this chair one afternoon Broad told me of his fears that the spirit world in the mid-twentieth century was losing all its colour. Not that spirits as such had finally gone to rest – new reports of hauntings, poltergeists and so on reached him every day – but it seemed that these modern spirits no longer cut the dash they used to do. Their activities were becoming – dare he say it? – increasingly vulgar. Only the previous day he had heard of a poltergeist which was shifting caravans around a holiday camp near Great Yarmouth. If this trend continued, Banquo would soon be advertising tartans on the television and Hamlet's father taking coach-parties round Elsinore. The old philosopher's face fell as he contemplated the disagreeable prospect, and I understood only too well why he had concluded his celebrated *Lectures on Psychical Research* with this remark: 'For my own part I should be more annoyed than surprised if I should find myself in some sense persisting immediately after the death of my present body.' I do hope he is not now passing his post-mortem retirement in the great caravan site in the sky.

But the evidence, in so far as it continues to reach us from the other side, is not encouraging. In fact – if evidence it is – Mr Playfair's account of the goings-on in an Enfield council house in 1977 suggests that the decline in standards which Broad detected twenty years ago has now become a rout. This is not a book for sensitive souls, nor is it a book about them.

The objection is not to the premises as such. Wood Lane, Enfield, Middlesex is probably a perfectly respectable locality. The fact that the ghost is not the owner-occupier, regrettable as it may be, is nothing new; and if a ghost chooses to be semi-detached, then that – as any student of structuralism will realise – has a certain logical consistency to it. But with ghosts as with people, it is not where you are but what you are which matters: and what the Enfield poltergeist is, is distinctly not nice.

Ghosts have of course a reputation for being horrible (even most horrible at times), but they have surely never before been so manifestly uncultivated, pettifogging or cheap. In the old days a ghost might be expected to give itself a proper introduction – 'I am the ghost of Christmas past' – but now in December 1977 the Enfield ghost shouts through the wall: 'I come from Durants Park, I am 72 years old . . . Shut the fucking door . . . I want some jazz music, now go and get me some, else I'll go barmy.' In the old days spiritual graffiti-writing was an art – 'Mene, mene, tekel, upharsin' written in letters of fire on the wall of the banqueting chamber – but now in Enfield the poltergeist, having laid out a row of potted cactuses on the kitchen floor, writes 'I am Fred' in black sticky-tape on the bathroom door. In the old days, if an offended spirit meddled with the cutlery it would pick out a dagger and wave it before the glazed eyes of the Thane of Cawdor, but now it takes a teaspoon from the dresser and bends it before the camera of a reporter from the *Daily Mirror*.

Still, if there is one thing a philosopher ought to be it is philosophic. And Broad was generous besides. Had he lived to hear of the Enfield poltergeist, I do not suppose it would have entered his head to regard bad manners on the part of a ghost as evidence of bad faith (or even of bad observation) on the part of the ghost-hunter. But here he and I part company. For, in the matter of ghosts, I have always subscribed to what may be called the Argument from Lack of Design. The argument simply stated is this: if it's ugly, formless or trivial it's probably fake.

I expect something super of the supernatural. If spirits exist we shall know them by their works – and their works *ex hypothesi* will be potent, elegant and grand. Even allowing for the occasional off-day, I cannot believe that any spirit worth his salt would challenge the laws of physics by bending teaspoons, or would challenge the will of living human beings by telling them to fuck off and get some jazz music. We live, I know, in a changing world, but not surely a world in which the entelechy which orders the supernatural universe is a cross between Fungus the Bogeyman and Mrs Thatcher.

Who, then, is Fred? And what is Mr Playfair doing writing a book about him? For an answer we may perhaps turn to Professor John Taylor, who has known Fred – or rather a mate of his called Uri – from both sides.

Eight years ago Uri Geller – one-time fashion model, conjurer and convicted con-man – hit our television screens with a miraculous display of spoon-bending. The effect on the British public was remarkable. There were those who wondered at Geller's supernatural powers, and those again who merely wondered about them. John Taylor, Professor of Mathematics at London University, did both by turns.

Taylor is an empiricist. Not for him the argument from good form or the lack of it, but rather the argument from galvanometers and strain-gauges. Confronted by the phenomenon of paranormal metal-bending, he was determined – teaspoons or no teaspoons – to get at the scientific truth: and when it turned out that the truth changed over the years, so much the better – for then Taylor could tell us about it in two books rather than one.

In his earlier book, *Superminds* (1975), he described how his research on the 'Geller effect' had at that time persuaded him of the reality of strange forces beyond the ken of physics. But in *Science and the Supernatural* (1980), he tells how further investigation of the world of psychical phenomena (extra-sensory perception, table-turning, poltergeists, divination and so on) has made him change his mind: 'We have searched for the supernatural and not found it. Only poor experimentation, shoddy theory and human gullibility have been encountered.' Years of research on the paranormal bending of metal have, it seems, revealed nothing more than the all too normal bending of facts.

Taylor, reincarnated, emerges as a stern and sometimes uncharitable critic of the psychical fraternity to which he previously belonged. And yet, like many an apostate, he continues to have a love–hate relationship with his former Church. Though he can no longer countenance its rites, neither can he bring himself to ignore them altogether. Habits of faith (and habits of writing bestsellers) die hard – and if there is a good story to be told, Taylor may be counted on to tell it, even when he no longer believes it to be true. Thus he begins with 'spontaneous human combustion', ends with survival after death, and leaves out precious little in between. In fact, so anxious is he not to miss a trick (in more senses than one), that at times he seems determined to overplay his adversaries' hand. It gives him as much satisfaction to unmask a straw ghost, to topple the battlements of a caravan site, or to return a teaspoon to its scabbard, as it does to find fault with a serious piece of laboratory research. The result is that the book as a whole reads more like campaigning journalism than like scholarship.

That is not to say that Taylor's tactics are unsubtle. Crusader though he is, he does not attempt to rush the enemy's camp in the naïve belief that at the cry of 'Science akbah!' the walls of superstition will immediately come tumbling down. Instead he contrives to lure the defenders out into the open before he cuts the ground from under them.

With a technique that is almost Socratic in its disingenuousness, Taylor presents his argument in the form of a running dialogue between three characters: the sheep, the wolf in sheep's clothing, and the wolf. First comes the sheep, a trusting layman who is ready to believe everything he hears. This likeable fellow, who behaves for all the world as if ectoplasm would not melt in his mouth, is given the job of setting up the psychic scene: 'There I was, minding my own business, when suddenly the man burst into flames ... The table rose from the floor before my very eyes ... The lady went into a trance and found herself reliving her earlier life as a servant to the King of France ... The spoon bent without the boy so much as touching it ...' etc. Next comes the liberal-minded physicist who on the face of it appears quite willing to believe it all, provided only that his measuring instruments can confirm that the events described have actually occurred: 'If the man caught fire, then there ought to have been some heat given off which would register on a thermometer ... If the table rose from the floor then a spring balance should indicate that temporarily the table had negative mass ... If the lady was alive eight hundred years ago then she ought to remember the eclipse of the sun in the year 1180 ... If the spoon bent without the boy touching it then it should not be covered with his fingerprints ...'. And finally comes the forensic scientist, a detective-cum-psychiatrist who knows all about fraud and something about Freud: 'The witness was later admitted to hospital suffering from conflagratory hallucinations . . . The table was lifted by a paid accomplice . . . The lady from France, though she would not admit it even to herself, had read it all up in an historical romance . . . The boy was caught in the act of bending the spoon behind his back . . .'. Alas for the loss of innocence! When you sup with Uri Geller you must, it seems, not only take a long spoon but keep a very careful eye on it.

Because Taylor attempts to cover so much, he does not always do the job effectively; and sometimes, indeed, he spoils his case against the psychical researchers by committing methodological errors of his own. Inevitably there will be readers who find his treatment shallow: a magical demystifying tour which is too much of a whistle-stop affair. There are certainly people (and I know several of them among my academic colleagues) who believe that when we are confronted by prima facie evidence of the paranormal we should lean over backwards to accept even the strangest claims until their dubiousness has been established beyond all reasonable doubt. Such people are bound to wish that Taylor had chosen to examine a few of the better-documented cases in more depth. But for my own part I confess my prejudice: life is too short, the natural world too interesting, for it to be insisted that we approach even the show-pieces of psychical research with an obsessively open mind. Though I still nurse a childlike hope that one day I myself shall meet with some grand and incontrovertible manifestation of the spirit world, I am not prepared to go looking for it on my hands and knees. For the moment the ball remains firmly in the supernatural's court. In Taylor's words, the paranormal has turned out upon investigation to be totally normal.

Or has it? Only, I think, if we accept a very odd definition of the normal. For when Taylor's book is finished, two major aspects of the subject remain unexplained. First, why do the pedlars of the paranormal do it? Is it really 'normal' for otherwise honest citizens to cheat, to lie, to fake the results of experiments, and to practise delusion – and self-delusion – on a scale which has almost become institutional? Second, why does society pay so much credulous attention to them? Is it really 'normal' for the rest of us to treat tricksters, hysterics, charlatans and practical jokers as if they were supernaturally inspired?

An answer to the second question would clearly go a long way to answering the first. For in one form or another, attention is precisely what most of those people who claim to possess or to have witnessed supernatural powers are seeking. It may be an Israeli conjurer who is simply after fame and money, it may be a newspaper reporter who is trying to impress his editor, a Cambridge research student who is hoping for a degree in parapsychology, or any of the other unsung men and women who are attempting to be bigger than they really are. It may even be, as in Enfield, a teenage daughter of a broken marriage who is seeking with her voices and her sticky-tape to advertise her own emotional distress ('Why do girls have periods?' asked the voice of the poltergeist in one of its more revealing moments). But whoever it is, whatever their particular motive, there can be no doubt that involvement with the supernatural is a sure way of gaining special recognition. The question is, why are the rest of us so ready and willing to give it?

We live in a culture where belief in the supernatural is strong and growing. In 1660 Joseph Glanvil (quoted by Taylor) could write: 'The present-day world treats all such stories with laughter and derision and is firmly convinced that they should be scorned as a waste of time and old wives' tales . . .'. But in 1977 a poll of American university professors showed that 65 per cent thought that the existence of paranormal phenomena was either established or extremely probable (the proportion of believers being, not surprisingly, considerably higher among engineers and physicists than among psychologists and anthropologists). To some extent, this growth is self-sustaining. For belief breeds belief: once we have abandoned the constraints of rationality, it becomes only too easy to prefer simple-to-grasp supernatural explanations to difficult-to-grasp natural ones.

Thus, a lady from Dublin dreams of the assassination of President Kennedy on the night before his actual death, and we accept it at once as a 'precognitive dream' – forgetting that every night of the year there are dreamed on this planet about 20,000 million dreams (five per night per head of the world population), making it almost inevitable that some of them come true. An astrologer discovers that in over half the school classes in England there are two or more children with exactly the same birthday, and we immediately detect some cosmic influence at work – forgetting that in any group of 23 people there is a 50 per cent probability that two will have the same birthday by chance. A gypsy fortune-teller reads our hands and tells us some secret known only to ourselves, and we attribute it at once to second sight – forgetting that while *we* were looking at our hands she was looking at our face and that the reading of facial expression is a skill which the human species has been slowly perfecting over at least the last five million years.

Once we believe, and want to go on believing, there is no reason to stop. Not only can we find confirmatory evidence all around us, but counter-evidence ceases to exist. For we can always give a topsy-turvy interpretation to anything which might otherwise suggest we are being fooled. When, for example, Janet Harper, the teenage girl in the Enfield council house, confesses under hypnosis that she and her sister are responsible for causing 'all the trouble', Mr Playfair is worried for only a moment before he decides that what Janet actually means is that she and her sister are *indirectly* responsible because it is they who have upset the poltergeist. When it appears that Stephen North, one of the child-star spoon-benders, always gets his best results just as Mrs North interrupts the experimental session by bringing in the tea, Professor John Hasted concludes that Stephen is marshalling his paranormal powers 'in response to an offer of reward', rather than that he is taking advantage of the tea-time confusion to cheat (see J. B. Hasted's paper to the third International Conference of the Society for Psychical Research). Or when, as so many of us must know, one of our friends tells us in hushed tones of his own encounter with a ghost, we treat the story as evidence of just how 'frank' our friend is being rather than as prima facie evidence that he is fibbing. But if such distortions of judgement are the consequences of belief, where does that belief begin? I think it begins with the fear of being seen to be an unbeliever.

In a culture where belief in the supernatural is commonplace, sceptics who publicly express their doubts are on a hiding to nothing. They will be disliked for being right, and disliked for being wrong. When they can prove their case, the best they can expect is to be branded as a know-all. When they cannot prove it, they are criticised for daring to question what others take on trust. They will be accused of intellectual conceit ('There are more things in heaven and earth than are dreamt of in your philosophy'); of insensitivity ('Man will not perish through want of wonders; only through want of wonder'); of sour grapes ('Just because it hasn't happened to you'); of protesting too much ('It has happened to you, only you're scared to admit it'); of failure to see the good in other people ('How dare you suggest he's a liar!'); of being no good themselves ('You wouldn't accuse him of lying, unless you too were versed in lying'); and so on. Worse still they may be accused of a kind of spiritual infanticide ('Every time a child says "I don't believe in fairies" there is a little fairy somewhere that falls down dead').

It is a rotten trick to knock down fairies. And yet, when all is said, perhaps it is the believers rather than the sceptics who are the greater criminals. For it is they who with their patent superstitions and quack theories drug us into a state of intellectual apathy in which we fail to attend to the true mystery, man. The strangest thing on earth is not ghosts or reincarnation or ESP; it is muddled, deluded and deluding *man*. It is Professor Broad, John Taylor, Uri Geller and Fred.

16

KARMA IS RAINING ON MY HEAD

Mind and Nature: A Necessary Unity
by Gregory Bateson

At the end of his life, the distinguished biologist C. H. Waddington took part in a discussion about the nature of mind. The circumstances were unusual. Waddington lay flat on his back, and his words were read from a prepared text by a friend. The discussion between him and the other two participants was lively: until, that is, there came a point when Waddington, having momentarily silenced his colleagues, abruptly left the room. The platform on which he was resting sank beneath him and his body was committed to the flames.

Gregory Bateson, if he had been present at the Edinburgh crematorium, would no doubt have felt sad at the death of his old friend. Yet he would surely have relished an event which in so many ways illustrated his own philosophical obsessions: his concern with birth, death, rites of passage, communication, meta-communication, mind, nature, and especially with the logical status of these things. Here, if there ever was one, was an example of a 'metalogue': a mind, in the presence of minds, discussing with other minds the nature of mind. And here was a fairground of conceptual paradox. Waddington was dead, and yet in both body *and* mind he was still present in the room. Waddington's life was at an end, and yet his 'karma' (Bateson's word) would continue to rain down like ash on the mourners as they left the crematorium and is raining on this page right now.

Bateson himself nearly died during the writing of this book. But, having survived a serious illness, he went on to produce what his publishers now call his major life's work. It is a work which, as Bateson readily admits in the last chapter, is far from finished; he has not, as Waddington had not, found final answers to the problems which perplex him. If it should happen that these two remarkable thinkers meet again, then I hope – for the sake of any of the rest of us who may join them in Elysium (and none, I imagine, would want to join a club which did not admit these two as members) – that they will continue the debate. Maybe, when no longer threatened by a deadline, they will make a better job of it than Bateson does in *Mind and Nature*.

Bateson's theme is that biological evolution is itself a *mental* process. That this theme is, as the publishers announce on the dust-jacket of the book, a 'startling' one, few would deny. But no less startling is the logical sleight-of-hand by which Bateson tries to persuade us to take the proposition seriously. His argument rests fair and square on the fallacy which logicians know as affirming the consequent. It runs like this. If something is to be called a mental process, it must have certain properties (which Bateson lists); evolution has these properties; therefore evolution is a mental process. In just the same way, one might argue: If it is Christmas Day, there will be no postal delivery; there is no postal delivery on Sunday; therefore Sunday is Christmas Day.

The properties which Bateson lists as 'criteria of mind' – a mind is an aggregate of interacting parts, mental process requires circular chains of determination, etc. – are uncontroversial and on the whole well chosen. To those readers who are not already thoroughly familiar with these mental properties, his discussion of them may at times prove illuminating. But when Bateson goes on to argue that because evolution – not to mention (as he does) life, ecology, the biosphere – also has these properties, evolution (life, ecology etc.) must be the same kind of thing as mind, it amounts at best to a mere verbal conceit. At worst, it amounts to a piece of moral blackmail. The strong implication of Bateson's argument (though he does not state it explicitly) is that if the whole of organic nature is imbued with qualities of mind, then we as human beings should treat nature with the respect due to a human mind. There are, I know, good reasons for not cutting down the Amazonian forests: but the idea that such destruction is equivalent to psychosurgery is not one of them.

There may well be real parallels to be drawn between the workings of mind (in the usual sense) and the workings of natural selection. Bateson is right to insist that concepts drawn from the behavioural sciences may provide useful metaphors for thinking about evolution. For example:

My theory may be described as an attempt to apply to the whole of evolution what we learned when we analysed the evolution from animal language to human language. And it consists of a certain *view of evolution* as a growing hierarchical system of plastic controls, and of a certain *view of organisms* as incorporating this growing hierarchical system of plastic controls. The Neo-Darwinist theory of evolution is assumed; but it is restated by pointing out that its 'mutations' may be interpreted as more or less accidental trial-and-error gambits, and 'natural selection' as one way of controlling them by error-elimination . . . Error-elimination may proceed either by the complete elimination of unsuccessful forms (the killing-off of unsuccessful forms by natural selection) or by the (tentative) evolution of controls which modify or suppress unsuccessful organs, or forms of behaviour, or hypotheses.

But Bateson is not right to present these ideas as if they were new ideas for which he himself can take the credit. The quotation above, which neatly summarises part of his argument, is not in fact from his own book but from Karl Popper's celebrated essay 'Of Clouds and Clocks', published fourteen years ago.[1]

The question of whether Bateson is saying anything new arises still more obviously when he comes to discuss the reverse side of his thesis, namely that mind is an *evolutionary* process. Intellectual progress, Bateson argues, depends like biological progress on selection from a pool of novel possibilities. But novelty must arise out of randomness: 'without the random there can be no new thing'. New ideas when they first arise have no direction, just as genetic mutations have no direction; only once they have come haphazardly into the world can they be tested for their fitness in terms of their coherence with the existing body of thought, and of their compatibility with the external environment.

Now, in case the reader should think that this idea of Bateson's has come haphazardly into the world, compare William James in 1880:

'I can easily show that throughout the whole extent of those mental departments which are highest . . . the new conceptions, emotions, and active tendencies which evolve are originally *produced* in the shape of random images, fancies, accidental outbirths of spontaneous variation in the functional activity of the excessively unstable human brain, which the outer environment simply confirms or refutes, preserves or destroys – selects, in short, just as it selects morphological and social variations due to molecular accidents of an analogous sort.'

Or Mach in 1895:

'The disclosure of new provinces of facts before unknown can only be brought about by accidental circumstances . . . From the teeming, swelling host of fancies which a free and high-flown imagination calls forth, suddenly that particular form arises to the light which harmonises perfectly with the ruling idea, mood or design. Then it is that that which has resulted slowly as the result of a gradual selection, appears as if it were the outcome of a deliberate act of creation.'

Or Souriau in 1881:

We know how the series of our thoughts must end, but not how it should begin. In this case it is evident that there is no way to begin except at random. Our mind takes up the first path that it finds open before it, perceives that it is a false route, retraces its steps and takes another direction . . . By a kind of artificial selection, we substantially perfect our own thought and make it more and more logical.

In an important essay on evolutionary epistemology, D. T. Campbell has listed twenty-six previous statements by different authors of the same idea.[2]

If it could be claimed that Bateson in this book had presented a set of old ideas with novel force or clarity, there would perhaps be some excuse. As he himself rightly says in a chapter on multiple versions of the world, 'Two descriptions are better than one', as in the case of binocular vision, where two eyes, seeing the same scene from slightly different points of view, add a third dimension to the image. But when the monocular version, such as Bateson's book provides, is, as I see it, frankly astigmatic, it is not obvious that it deepens our perception of the issues.

The text of the book is so unfocused, and many of the ideas so cock-eyed, that it requires from the reader immense (and sometimes fruitless) effort to see what it all means. While apparently writing for a general audience, Bateson employs technical jargon with an abandon I have seldom encountered even in the densest scientific prose. He uses familiar words in strange places ('the elephant . . . is *addicted* to the size that it is'), and strange words (*pleromatic*, *exoteric*) in places where familiar ones would do. He delights in archaisms (*saw* for a wise saying, *atomy* for atom) and in ugly neologisms (*creatural*, *characterological*, *stochasticism*). And he deliberately lays verbal trip-wires for the reader: 'Science, like art, religion, commerce, warfare, and even sleep, is based on presuppositions.' Even *sleep*? The shock, I suppose, of non-recognition.

But one of the risks a writer takes in making his reader work so hard is that the reader will keep working even when the writer has not adequately prepared the ground. If Bateson's text says 'The buzzer circuit (see Figure 3) is so rigged that current will pass around the circuit when the armature makes contact with the electrode at *A*', an attentive reader is surely entitled to be puzzled when he cannot find an electrode marked *A* in the figure. If the text says, in talking of acoustic beats, 'The phenomenon is explained by mapping onto simple arithmetic, according to the rule that if one note produces a peak in every *n* time units and the other has a peak in every *m* time units, then the combination will produce a *beat* in every $m \times n$ units when the peaks coincide', the reader is entitled to think twice about $m \times n$ and to conclude that Bateson has simply got it wrong. And once the reader has realised that he should distrust Bateson's circuit diagrams and his arithmetic, he may be alerted to the fact that he should also distrust Bateson's science.

Perhaps it would be pedantic of a zoologist to object to the statement that 'salmon inevitably die when they reproduce' (after all, a fair number of female salmon *do* die after spawning). But a psychologist will justifiably object to the pseudo-scientific statement: 'With the dominant [left] hemisphere [of the brain], we can regard such a thing as a flag as a sort of name of the country or organisation that it represents. But the right hemisphere does not draw this distinction and regards the flag as sacramentally identical with what it represents.'

Still, in the midst of this pretentious and muddled book Bateson occasionally reveals a disarming humility. 'Epistemology', he says, 'is always and inevitably *personal* . . . What is *my* answer to the question of the nature of knowing? I surrender to the belief that my knowing is a small part of a wider integrated knowing that knits the entire biosphere or creation.' It's all right, Bateson, you can put your hands down. No need to surrender. That particular belief has no fire-power in it whatsoever.

17

WHAT IS MIND? NO MATTER
WHAT IS MATTER? NEVER MIND

The Mind's I: Fantasies and Reflections on Self and Soul
edited by Douglas Hofstadter and Daniel Dennett

Both this and the following statement are false. *The Mind's I* is well worth £9.95.

One does not have to be a philosopher to realise that I have already gone further than a humble book-reviewer should. The value of the book under review is, you might think, a matter of opinion. Not so. It has now been established as a logically necessary truth.

Consider the first statement in the paragraph above: 'Both this and the following statement are false.' If this statement were true, it would by its own claim be false. Therefore it cannot be true, and the only logical alternative is that it is false. Consider the second statement: '*The Mind's I* is well worth £9.95.' If this statement were also false, then both statements would be false – which is precisely what the first statement claims. But that would mean that the first statement was true after all, which we have just established that it cannot be. Therefore the second statement cannot be false, and the only logical alternative is that it is true.

People who like this kind of thing will undoubtedly find this book the kind of thing they like. Counting myself among them, I would reckon *The Mind's I* cheap at twice the price.

The Mind's I – The Mind *is* I, The I *of* the Mind? – is a set of essays, fairy stories and brain-teasers by nineteen different authors, chosen by Hofstadter and Dennett because they illuminate in one way or another the problems of self-reference, personal identity, consciousness, and the relations between language, mind and brain. The editors have, as they put it, 'arranged and composed' the pieces, and they have provided for each an engaging 'reflection' or commentary. Although the book is philosophical in the best sense of the term, there is little in it by way of straight philosophy; and although it touches on issues central to psychology, neurophysiology and the computer sciences, it assumes no technical or factual expertise.

'In poetry', Auden wrote in *The Dyer's Hand*, 'all facts and all beliefs cease to be true or false and become interesting possibilities.' *The Mind's I* itself comes close to poetry. It is a book of fantasies and thought-experiments, an exploration of possible worlds, possible people, possible machines and possible divinities. True or false – or, as in several cases, neither one thing nor the other – these possibilities are designed to make us consider and reconsider a host of concepts which we take for granted.

Once upon some future time, imagine your brain is physically separated from the rest of your body, but with all the usual connections to muscles and sense-organs preserved by radio links. If your brain is, say, in a bath of warm saline in a hospital in Paris, while your body is walking the streets of London, where will *you* be? If your body then commits a felony, who or what should go to prison? Imagine a machine which can make a perfect copy of any existing physical structure, atom for atom. Suppose it makes a copy of you. What will you say to this copy when the two of you are introduced (and what will the copy say to you)? Suppose, having taken all your vital statistics in 1982, the machine does not produce the finished copy until 1992. What will you say to yourself in ten years' time, when you meet yourself as you were ten years ago? Imagine the nerve-cells in your head are one by one replaced by microchips, which function exactly like the original nerve-cells. At what point, if any, in this process of replacement will your stream of consciousness be interrupted? Suppose that instead of being put into your head, the replacement chips are kept in a laboratory from which connections into the system are made by remote control. Suppose, indeed, they are kept in lots of different laboratories. What will it feel like when your replacement brain has thus been spread around the world? Imagine you say to God, 'God, I am afraid of sinning, and I know I shall regret it', and He says: 'That's all right then, I won't stop you sinning, but I'll take away your feelings of regret.' How will you respond to this well-meaning offer? Imagine you wake up one morning and find yourself still in the middle of a dream . . .

These and other still more worrying paradoxes are explored here in the form of literary fables, serious if playful academic papers, and reviews (in the case of Stanislaw Lem's 'Non serviam', a review of a non-existent book). Some of the pieces, such as Borges's 'Borges and I' and Turing's 'Computing Machinery and Intelligence', are classics. Others have been previously published but are less well known. And others appear to have been specially written for this book. Dennett has included among other things Hofstadter's 'Prelude . . . Ant Fugue' taken from the latter's *Gödel, Escher, Bach*, and Hofstadter has included Dennett's 'Where am I?' taken from the latter's *Brainstorms*. Besides their joint and several contributions to the commentaries and introduction, the editors have also provided a helpful guide to further reading. (Alas, this guide is not quite as helpful as it seems. A paper, cited as Lenneberg 1975, turns out to be no more than a summary by someone else of an unpublished experiment, which adds nothing to the summary already given in this book. Eric Lenneberg committed suicide in 1975. Perhaps Hofstadter, who at one point in *The Mind's I* conducts a conversation with Einstein's brain after Einstein's death, knows something which the rest of us do not.)

You might think that any philosopher who valued his academic reputation would have been shy of putting his name to a collection which is so unashamedly exotic and bizarre. Indeed, in *Brainstorms*, published only three years ago, Dennett's strange story about losing his brain (or was it his body?) was left to the last chapter, where it was described somewhat coyly as a 'dessert'. But here such a disclaimer would be out of place. For the whole book is a book of desserts, without any attempt to provide a well-balanced or more wholesome diet. Nanny, of course, will tell you that such rich fare cannot be good for you: too much of the Tao, not enough of the oaTs.

I can't help feeling that Nanny might be right. This *is* a rich book, and a dangerously unbalanced one. Paradoxes are fun; they can be illuminating. But we should be wary of the temptation to celebrate paradoxes as a royal road to some higher level of reality – as if, with their help, we might expect to discover a conceptual world where contraries are all resolved, where 'Is' and 'Is not', 'So' and 'Not so', are smoothed away. Such Taoist doctrines may have proved productive in other fields of science – in atomic physics, for example, where irreconcilable notions about the nature of matter have apparently been resolved by the discovery of quantum field theory – but they are not what is most obviously needed in the science of mind.

Though one might conclude from this book that what is needed is a revolutionary new theory of the relation between mind and matter, I am not persuaded that many of the paradoxes presented here represent mysteries which can only be resolved by an appeal to principles as yet unknown. As often as not, they seem to represent nothing more than good old-fashioned muddles, of the kind which human thinkers habitually get into when they carry over their familiar concepts into unfamiliar territory. At the risk of being accused of teaching philosophers to suck eggs, I would suggest that what is missing from this book is the recognition that our ideas and our language operate – and are meant to operate – only within the limits of the 'game' for which they are designed. Human perception, human thought and even human ways of doing philosophy have evolved as adaptations to the world of the experienced past, not of the yet-to-be-experienced future. It is the world of the past which constitutes what John Bowlby has called 'the environment of adaptiveness'. And, as Alice discovered when she observed the croquet game in Wonderland, *outside* the environment of adaptiveness anything can happen.

Examples abound, not merely in fantasies and thought-experiments, but in well-tried fact. In normal circumstances, people have, for example, no trouble making an unambiguous decision about whether an object is coloured red or green. But let someone be induced by an experimental psychologist to look at a red object with one eye while he looks at a green object with the other, and he will – when conditions are right – confidently report that what he sees is a single object which is coloured 'red and green all over' (not part red part green, not red and green by turns, but red and green all over). In normal circumstances, people have no trouble making sensible three-dimensional interpretations of flat line-drawings. But let someone be shown the drawing in Figure 13, in which M. C. Escher has cunningly flouted the familiar rules of linear perspective, and what he will see is an 'impossible' staircase which goes on and on for ever. In normal circumstances, people have no trouble using the common or garden ideas of 'truth' and 'falsity' for the analysis of human discourse. But let someone be persuaded by a Cretan philosopher into making the (wholly unlikely) statement 'I

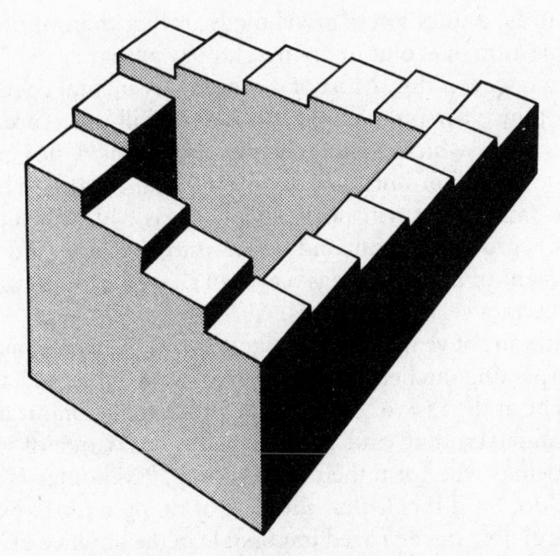

Figure 13. Escher's impossible staircase.

am lying', and he will find, whatever his mental state, that he has made a statement which he never meant to make.

They are fun, such paradoxes. But surely the last thing we need to do is to *resolve* them: to waste our philosophical energies on, say, inventing a new definition of colour, or a new kind of 4-D 3-D space, or even (*pace* Bertrand Russell) a new typology of 'truth'.

And the same goes, I think, for most of the paradoxes of mind and matter which are at the centre of this book. What the book so cleverly demonstrates is that when people carry over their common-sense concepts about mind into the Wonderland of Science Fiction computers, teleclones and brain surgery, they can and do get those common-sense concepts in a twist. They find themselves concluding, for example, that a person can be both here and there at the same time, or that consciousness can reside in the pages of a book, or that a machine can have free will. But the question which is often implicit in the stories and the commentaries – namely, 'Can we invent a new kind of philosophical elastic which will save us in future from the embarrassment of having our mental concepts dangling round our knees?' – seems to me the wrong one. The question ought to be 'Can we use this embarrassment to discover just what the limits on our existing concepts are – and why?' In other words, a question of psychology, rather than of philosophy. And a question of evolutionary psychology at that.

My own view is this. Most of our everyday mental concepts – the ideas of mind, person, consciousness, free will etc. – are primitive concepts with which human beings are saddled not merely by cultural convention but by a long biological history. They are, if you like (and some will not), 'innate ideas': ideas which people inevitably grow up with and grow into, because the cognitive development of human beings has been shaped by natural selection to meet certain very special ends.

For millions of years the most significant and at the same time the most perplexing intellectual task that all human beings have had to face has been the task of doing what I have called 'natural psychology' – understanding and predicting the behaviour of the fellow human beings who form their social group. Psychology is a difficult thing to do. So difficult that the task of doing natural psychology would probably have proved impossible in the absence of some sort of conceptual guidelines (witness, by comparison, the failure in recent times of open-ended systems of academic psychology such as behaviourism). But human beings in the course of evolution have in fact developed their own guidelines. And they have done so by a remarkable device: the invention of the faculty of reflexive consciousness. Reflexive consciousness, by giving each individual a picture of the 'psychological structure' which underlies his own behaviour, provides him with a framework for interpreting the behaviour of others like himself.

Thus most of our everyday mental concepts are, in this sense, natural psychological concepts – which we can no more do without or wish away than we can wish away our human concepts of colour, space or truth.

Dennett, in his writings elsewhere, sometimes seems close to adopting a similar perspective. In this book, however, considerations of the evolutionary function of human mental concepts receive no quarter. Dennett joins Hofstadter in Wonderland, where like a couple of Mad Hatters they invite us to take tea at a table miles away from the environment to which our concepts are adapted. And, teasing though their paradoxes are, the questions they raise – Can machines think? What would it feel like if my head was separated from my body? Whose side is God on? – are ultimately silly questions. About as silly as a former Archbishop's question 'Is AID *adultery*?', or President Reagan's 'Can the US *win* a nuclear war?' No, of course machines can't think. The environment of adaptiveness of the term 'think' is the world of human intelligence, just as the environment of adaptiveness of the term 'adultery' is the world of human sexual intercourse, and the environment of adaptiveness of the term 'winning a war' is the world of conventional military strategy.

I don't say that Hofstadter and Dennett's questions will never be worth asking. One day their interesting possibilities might indeed become commonplace realities, and if that happens we shall have to re-adapt to this new world. But first things first. Both this and the following statement are false. Mental philosophy should not try to run before evolutionary psychology can walk.

FOUR MINUTES TO MIDNIGHT

In my deepest soul I hug the supposition that with God's 'Let there be', which summoned the cosmos out of nothing, and with the generation of life from the inorganic, it was man who was ultimately intended, and that with him a great experiment is initiated, the failure of which because of man's guilt would be the failure of creation itself, amounting to its refutation. Whether that be so or not, it would be as well for man to behave as if it were so.

Thomas Mann

18
AN IMMODEST PROPOSAL

At the University of Bologna in the thirteenth century there was a professor, Novella d'Andrea, who, it's said, was so distractingly beautiful that she had to deliver her lectures from behind a curtain. I have been thinking about Novella d'Andrea . . . thinking that what is sauce for Bologna is sauce for Cambridge. If Professor d'Andrea could do it, why not I?

Admittedly my reputation as a beauty is – or has been until now – rather less than Novella d'Andrea's. When in the past I have brazenly given my own lectures in open view of the students, full-frontal as it were, no one has swooned, stabbed themselves through the heart with their biros or wept passionate tears into their lecture notes. But I now realise where I have gone wrong: these past performances must have left far too little to the imagination.

Next term I too shall give my lectures from behind a curtain. And by way of explanation I shall put it abroad that Dr Humphrey is so handsome that, regretfully, he has felt obliged to hide himself from public view. I shall add for good measure that Dr Humphrey's voice is so seductive and the content of his lectures so disturbingly profound that he has thought it best that he should not actually speak out loud. And then, silent and invisible before the massed ranks of students, I shall wait for the myth-makers to do their work. By the end of the year I fully expect to be the most celebrated lecturer in Cambridge.

The principle is not a new one. Indeed, I imagine it has been exploited by enterprising image-salesmen since time began. Film stars, diplomats, poker players, politicians, all know – if they know anything – that if you want people to believe that you are what you are not, then you must not show them what you are. If Ronald Reagan is to be thought statesmanlike, he must not allow himself to speak without a script. If Brigitte Bardot is to be thought young, she must not allow herself to appear without her make-up on. Or – and here is a trick which even Mr Reagan's PR men have hesitated to employ – if an ape is to be thought to be a Don Juan of the forest, it must not be seen to have a penis merely the size of a man's thumb. The carcase of the first gorilla brought to England a hundred years ago had its genital organs deliberately cut off so that the inflated expectations of the public should not be disappointed by the truth. Out of sight, but – we must assume – not wholly out of mind.

But do not get me wrong. I am not suggesting that this kind of humbug is necessarily objectionable. It may do us good to imagine our heroes and heroines to be bigger, more glorious, less human than they really are. And we undermine these illusions at our peril. Jonathan Swift in his poem *Cassinus and Peter* tells the cautionary tale of a young man who, believing his mistress to be free of ordinary mortal cares, followed her into the privy to check that she was not, in fact, as human as the rest of us – and caused himself nothing but distress.

> No wonder that I lose my wits:
> Oh! Caelia, Caelia, Caelia shits.

A lamentable verse, to mark a lamentable discovery.

No, it is not always wrong to live with a false picture of others. But what *is* always wrong is to live with a false picture of oneself. The real danger is that Novella d'Andrea behind her curtain, Brigitte Bardot behind her make-up, or Reagan behind his publicity machine will themselves come to think that the fictions which they put about are true.

And that brings me to my point: a serious message for Mrs Thatcher, Chairman Andropov – and such others as succeed them. It's this. Whatever the public image you present, it is time you remembered privately that you are in fact no more worthy, no less human than the rest of us. Despite your offices, despite your clothes and titles, it is time you remembered that you too have to shit. And while you're there, remember St Augustine's observation, as true of you as it is of any other: *Inter urinas et faeces nascimur*. We are all born between urine and faeces. Lecture us from behind a curtain if you will, but do not put a curtain in front of your own mirror: because if you do, you may forget that you are human.

In 1957 at the Labour Party's debate on disarmament, Aneurin Bevan declared that he was not prepared to 'go naked into the conference chamber'. It is a phrase which has been echoed by Tory and Labour defence spokesmen alike; something similar was said at the Liberal Party conference in September 1981. But what was it that

Bevan had to hide? Bevan came into the world naked, and naked he left it. Why should he have been afraid to go naked into the conference chamber to discuss matters of global life and death? What he had to hide, as much from himself as from his adversaries, was nothing less than his humanity.

Of course, by the rules of the game he had to hide it. For no naked human being, conscious of his own essential ordinariness, the chair-seat pressing against his buttocks, his toes wriggling beneath the conference table, his penis hanging limply a few feet from Mr Andropov's, could possibly play the game of international politics and barter like a god with the lives of millions of his fellow men. No naked human being could threaten to press the nuclear button.

So I come to my proposal. Our leaders must be given no choice but to go naked into the conference chamber. At the United Nations General Assembly, at the Geneva disarmament negotiations, at the next summit in Moscow or in Washington, there shall be a notice pinned to the door: 'Reality gate. Human beings only beyond this point. NO CLOTHES.' And then, as the erstwhile iron maiden takes her place beside the erstwhile bionic commissar, it may dawn on them that neither she nor he is made of iron or steel, but rather of a warmer, softer and much more magical material, flesh and blood. Perhaps as Mr Andropov looks at his navel and realises that he, like the rest of us, was once joined from there to a proud and aching mother, as Mrs Thatcher feels the table-cloth tickling her belly, they will start to laugh at their pretensions to be superhuman rulers of the lives of others. If they do not actually make love they will, at least, barely be capable of making war.

19

FOUR MINUTES TO MIDNIGHT

In November 1945 Jacob Bronowski went as a member of the British Mission to the Japanese city of Nagasaki. In August that year President Truman, with the agreement of Winston Churchill, had ordered that the city and its population be destroyed by an atomic bomb. The bomb dropped on Nagasaki on 9 August killed 70,000 people. The bomb dropped on Hiroshima three days earlier killed 140,000. In the central square mile of each city nine out of every ten people died. Nine out of ten of those nine out of ten were not soldiers or politicians: they were children, mothers, grey-haired old men and women.

At the outbreak of war in 1939 such an attack by the Allies on non-combatant civilians would have been unthinkable. Civilised countries still clung to a morality which enjoined them to respect life and to limit suffering even of those in arms against them – a morality which taught a naval captain that having sunk an enemy ship he must rescue the survivors from the sea, and a prison-camp commander that he must treat with chivalry his prisoners of war. It was a morality which forbade at all times the use of indiscriminate violence against an unarmed population. The Allies had gone to war to defend these very values against the barbarous State of Nazi Germany.

But the world – and civilisation – had come a long way between 1939 and 1945. Hitler was dead. He had lost the battle. But the policies of terror which Hitler himself had pioneered had, it seemed, won the war.

Bronowski recalls that as he stood among the ruins of Nagasaki his imagination was dwarfed by the catastrophe. He was a man – we know – not often lost for words; but here he had none that were adequate. It was, he says, 'a universal moment . . . civilisation face to face with its own consequences'.[1]

The world had come a long way between 1939 and 1945. It has come far further since 1945. There are today in readiness for military use not two but 50,000 nuclear weapons, with a combined explosive power equal to more than a million Hiroshima bombs, or the equivalent of more than three tons of TNT for every individual on the earth. If these weapons should be used, then, as President Carter said in his farewell address to the American nation, 'more destructive power than in all of the Second World War would be unleashed every second for the long afternoon it would take for all the missiles and bombs to fall . . . A Second World War every second – more people killed in the first few hours than in all the wars of history put together.' Carter went on: 'It is now only a matter of time before madness, desperation, greed or miscalculation let loose this terrible force.'[2]

Bronowski made a television series, *The Ascent of Man*. Must what goes up come down? Looking at our progress over the last thirty years, it is hard to avoid the conclusion that mankind, having flown too near the sun, is already in a stall.

There are voices enough now raised in warning, the voices of statesmen who at other times have been attended to. Lord Mountbatten, speaking in Strasbourg a few weeks before he was assassinated: 'The world now stands on the brink of the final abyss . . .'.[3] Lord Zuckerman in a lecture to the American Academy of Sciences: 'The world has without doubt become a more perilous place than it has ever been in human history . . . The progress of the nuclear race clearly has no logic . . .'.[4] Professor George Kennan, former US ambassador to Russia, speaking in Washington in 1981: 'We have gone on piling weapon upon weapon, missile upon missile . . . like the victims of some sort of hypnotism, like men in a dream, like lemmings heading for the sea . . .'.[5] On the front cover of the *Bulletin of the Atomic Scientists*, the doomsday clock, set at ten minutes to midnight some years ago, was advanced by six minutes last January: four minutes to go.

I want to ask a simple question: Why? Why do we behave like lemmings? Why do we let it happen? In the words of Lord Mountbatten: 'How can we stand by and do nothing to prevent the destruction of our world?'

Mountbatten said in the same speech: 'Do the frightening facts about the arms race, which show that we are rushing headlong towards a precipice, make any of those responsible for this disastrous course pull themselves together and reach for the brakes? The answer is "no" . . .'. I want to ask how the answer can be 'no'.

I leave it to others to explain our extraordinary plight in terms of military necessity, economic competition, or the politics of the cold war. My concern is more primitive. I am concerned as a psychologist with the feelings, perceptions and motives of *individual* human beings. When a lemming runs, it is not pushed or pulled by outside forces, it runs to destruction on its own four feet. It is as individuals that we can and might apply the brakes, and as individuals that we can and do fail. Responsibility for 'this disastrous course' begins right here.

Perhaps there is an obvious answer, which is that we are simply unaware. Is it possible that we either do not know or else discount the dangers of the arms race? That we think the bonfire which is being built around us will never catch light – indeed that the larger it grows the less dangerous it becomes?

When I was a child we had an old pet tortoise we called Ajax. One autumn Ajax, looking for a winter home, crawled unnoticed into the pile of wood and bracken my father was making for Guy Fawkes Day. As days passed and more and more pieces of tinder were added to the pile, Ajax must have felt more and more secure; every day he was getting greater protection from the frost and rain. On 5 November bonfire and tortoise were reduced to ashes. Are there some of *us* who still believe that the piling up of weapon upon weapon *adds* to our security – that the dangers are nothing compared to the assurance they provide?

Yes, there are some of us. And it is hardly surprising that there are. For those in authority do little if anything to inform us of the dangers. We do not hear the British Prime Minister talking about the world 'standing on the brink of the abyss'. We do not hear the Defence Secretary talking about the nuclear arms race as being 'clearly without logic'. The Director General of the BBC protects television audiences from seeing the film *The War Game*, because he calculates, quite rightly, that people would find it alarming and distressing. Newspaper editors and defence correspondents have become apologists for official policy, instead of serving their traditional and honoured role as critics. And when we do get news of what's going on, it is couched in a language designed to enlist our admiration for the marvels of military technology and to quiet our fears and blunt our sensitivities. News-speak – Pentagon-speak – is employed by military spokesmen to distance us from the reality. It has become commonplace, as the Archbishop of Canterbury observed in a speech in Washington, to refer to the destruction of a city and its people as 'demographic targeting'.[6]

And yet . . . and yet not everyone, not even the majority of the population, is taken in. Opinion polls carried out in the past year show that, despite all the talk about the effectiveness of a deterrent strategy, nearly half the adult population expects nuclear war within their lifetime. Despite all the talk about civil defence, less than one in ten believes that they and their families would not be killed. And that of course is only the adult population; no one has yet carried out a national survey of our schoolchildren, but parents and teachers know that children too, and perhaps children above all, are deeply troubled. A County Inspector of Schools writes in a letter to *The Times*: 'I have sat in on discussion lessons when children have brought up the question of the Bomb. Many have come to accept that they may not live out their lives in full . . . Some smile about it . . . Others are most painfully aware of what is involved.'[7] Never in recent times, not since the plagues and famines of the Middle Ages, can so many people of this country have had such a pessimistic vision of the future.

But does this pessimism stir them into action? When questioned on behalf of the magazine *New Society*, 70 per cent of the sample said they were worried about nuclear weapons – but nine out of ten of this 70 per cent stated either that nothing could be done or else that they were unwilling to do anything. And even for the one in ten who said they might do something, the actions mentioned would seem to be totally incommensurate with the perceived dangers: they would go on a march, they would write a letter to the newspapers . . .[8]

It is as though we have become passive, fascinated spectators of the slowly unfolding nuclear Tragedy. I was taught at school that the essential quality of a Tragic Play is this: when the curtain rises you see a gun on the wall, and you know that in the last act the hero or heroine will take the gun from the wall and shoot themselves. It has to be so, the internal logic of the play allows no other ending . . . But now we are not the *audience* to this play. It is *we* who will get shot.

It is easy for those of us who are not historians to kid ourselves that nothing like this has ever happened before – and that because it has never happened before, it cannot really be happening now. But it *has* happened before, if never on such a disastrous scale. There are in fact dreadful precedents in history: times when whole groups of human beings, men and women whose love of life was no less than our own, have gone almost without protest to destruction – 'like the victims of some sort of hypnotism, like men in a dream'. I think of the long-suffering European Jews in the last war; of the way so many of them patiently took the trains to the extermination camps; of what happened in 1942 in the ravine near Kiev known as Babi Yar, where thousand upon thousand queued up for execution, mothers and fathers hand in hand with children, shuffling their way slowly forwards till they reached the front of the line and were gunned down. I think of the victims of the Stalinist purges in the Russia of the 1930s; of the way, week by week, people saw their comrades disappear into the torture chambers and the jails; they knew they would be next – and waited.

In her brave memoir of the purges, *Hope Against Hope*, Nadezhda Mandelstam, widow of the Russian poet Osip Mandelstam, describes how with disbelief she watched first her friends and finally her husband go the way of all the others.

> Later [she writes] I often wondered whether it is right to scream when you are being beaten and trampled under foot . . . I decided it is better to scream. This pitiful sound . . . is a concentrated expression of the last vestige of human dignity . . . By his screams a man asserts his right to live, sends a message to the outside world demanding help and calling for resistance. If nothing else is left, one must scream. Silence is the real crime against humanity.[9]

Why do we not scream? Why, when faced with the nuclear threat, do so many of us adopt a policy of quietism and collaboration? Why do we choose appeasement rather than protest? 'Daddy, what are *YOU* doing to stop the next war?'

That man is many different people – and the answer appropriate to one person will not necessarily be appropriate to another. But he is *me*, and he is *you*. And in trying to explain why it is that so many people are doing nothing, I will not take the easy path of suggesting explanations of a kind which always seem to fit *other* people so much better than ourselves – explanations in terms, say, of moral shallowness, unthinking obedience to authority, or plain stupidity. No doubt the world does have its fair share of mindless sheep, and no doubt they are doing nothing. But they are not alone. And I want here to bring the discussion nearer home: to focus on things I have felt in myself, which I know among my friends, and which I believe you too will recognise. In all of us there are powerful inhibitory forces working, which block or deflect us from effective protest. I shall speak first of Incomprehension and Denial, second of Social Embarrassment, third of Helplessness, and fourth, perhaps most sinister, of what I would call the Strangelove Syndrome – latent feelings of admiration, almost of appetite, for the Bomb and the final solution it provides.

I start with Incomprehension, where I suspect many of us both begin and end. Nuclear weapons are *not comprehensible*: neither you nor I have any hope of understanding just what they are and what they do. In saying that, I mean to belittle none of us; it is almost a compliment. For I do not see how any human being whose intelligence and sensitivities have been shaped by traditional facts and values could possibly understand the nature of these unnatural, other-worldly weapons. So-called 'facts' about the Bomb are not facts in the ordinary sense at all: they are not facts we can relate to, get our minds round. Mere numbers, words.

Let me repeat a fact. The Bomb which was dropped on Hiroshima killed 140,000 people. The uranium it contained weighed about twenty-five pounds; it would have packed into a cricket-ball. 140,000 people is about equal to the total population of Cambridge.

I, for one, cannot grasp that kind of fact. I cannot make the connection between a cricket-ball and the deaths of everyone who lives in Cambridge. I cannot picture the 140,000 bodies, let alone feel sympathy for each individual as he or she died. And when someone tells me – and I tell you – that a war between the United States and Russia will now mean a Second World War every second, and that the equivalent of 5,000 Hiroshima bombs will land in England, my imagination draws a blank. It is not just that I cannot *bear* the thought: I cannot even *have* the thought of 5,000 Hiroshima bombs . . . 5,000 times 140,000 equals 700 million: 700 million dead out of a population of 50 millions. Something wrong somewhere. Everyone getting killed ten or twenty times over . . .

We close off from such nonsense. Try as we may, we shall not get the message. Our minds are finely tuned by culture and by evolution to respond to the frequencies of the real world. And when a message comes through on an alien wavelength it sets up no vibrations. The so-called facts pass clean through us and away, like radio emissions from the stars.

There are strange and interesting precedents in history. When Captain Cook's great ship, *Endeavour*, sailed 200 years ago into Botany Bay, the Australian aborigines who were fishing off the shore showed *no* reaction. 'The ship' – I quote from Joseph Banks's journal of the voyage – 'The ship passed within a quarter of a mile of them and yet they scarce lifted their eyes from their employment . . . expressed neither surprise nor concern.' In the experience of these people nothing so monstrous had ever been seen upon the surface of the waters – and now it seems they could not see it when it came.

But theirs was a selective blindness. Cook put down his rowing-boats: *now* the natives were alarmed, now they looked to their defences. Blind to the greater but incomprehensible terror, they reacted quick enough to a threat which came within their ken.

We too react, selectively, to man-sized threats. It is not giant dangers or giant tragedies, but the plight of single human beings which troubles us. In a week when 3,000 people are killed in an earthquake in Iran, a lone boy falls down a well-shaft in Italy – and the whole world grieves. Six million Jews are put to death in Hitler's Germany, and it is Anne Frank trembling in her garret who remains stamped into our memory.

The story of Hiroshima too can be told as the story of individual human beings. The tale, for example, of a little girl:

> When my grandmother came back, I asked 'Where's Mother?' 'I brought her on my back,' she answered. I was very happy and shouted 'Mama!' But when I looked closely, I saw she was only carrying a rucksack. I was disappointed . . . Then my grandmother put the rucksack down and took out of it some bones . . . I miss my mother very much.[10]

Keiko Sasaki and her mother. But multiply the tragedy a hundred thousand times, and it no longer has any meaning to us. We are each too human to understand the killing-power of nuclear weapons, each too close to the good earth to understand how a metal cricket-ball could explode with the force of 10,000 tons of TNT. Each of us aboriginally blind.

We must live with this blindness. It will not change. I do not expect my dog to learn to read *The Times*, and I do not expect myself or any other human being to learn the meaning of nuclear war, or to speak rationally about megadeaths or megatonnes of TNT. The most we can ask for is an open recognition that neither we, when we protest against nuclear armaments, nor the generals and the politicians when they defend them, know what we are talking about.

And yet we do know *something* about what we are talking about. We know, if nothing else, that we are talking about something which would if it happened be very, very bad. And in the face of this knowledge we may find ourselves suffering from another kind of blindness, a blindness equally human but in many ways less innocent than the blindness which comes from lack of understanding. I mean the deliberate blindness which comes over us when we see something and then reject it: when we recognise the truth, or at least part of the truth, and – finding it perhaps too painful or inconvenient – we censor its access to our conscious mind. I mean what psychologists have called Denial. Call it wishful thinking if you like – or call it optimism, or the good old British habit of not taking things too seriously. It comes to the same thing.

There are of course obvious and good excuses for denial. It not only makes for a comfortable life; it makes, some would argue, for the only kind of worthwhile life there is. Certainly we cannot carry on as normal under the shadow of the Bomb. The prospect of nuclear war would, *if we allowed it to*, be totally distracting and totally depressing. It would, *if we allowed it to*, take away the meaning from the rest of our life and finish us off as creative and productive people. It is a prospect which flatly contradicts every other prospect we hold dear.

Human beings strive for consistency in their affairs. They cannot – at least they cannot for long – hold incompatible beliefs. Either, it seems, we look to the right of the picture or else to the left. Either we believe the world is threatened by extinction, or else we don't. How can we at one and the same time declare ourselves for human rights, devote ourselves to our children, labour to produce lasting works of art and scholarship . . . *and* take seriously a vision of the future in which there are no children, in which our books will never be read, and our paintings, our houses, our flower-gardens will end as dust? One or the other vision has to go.

Let us not be deceived. The dangers will not be diminished because we close our eyes to them. If we cannot carry on as normal under the shadow of the Bomb, then for the time being we have a duty *not* to carry on as normal. We live at a time when to deny the prospect of death may well cost us our lives.

Yet try telling that to other people. Try telling it in Gath and publishing it in the streets of Askelon . . . for the attitude of the daughters of the Philistines is little changed. To speak the truth among people who do not want to hear it is considered almost an aggressive act – an invasion of privacy, a trespass into someone else's space. Not nice, not done . . .

A year ago in Pennsylvania, USA, eight nuclear protesters who called themselves the Plowshares 8 (they included two priests, a nun, a lawyer and a professor of history) broke into a weapons factory and damaged the nose-cone of a Mark 12 missile with a hammer.[11] They were accused at their trial of burglary, criminal conspiracy and trespass. Each faced a maximum sentence of twenty-five years. 'This time', the prosecution said, 'you've gone too far.' They *had* gone too far, they *had* trespassed: but their trespass was not so much against the property of the General Electric Company as against the minds of the American public. By attacking the missile with a hammer they were forcing other people to think, however briefly, about a subject regarded as indecent – the question of just what such a missile might be for.

Their behaviour was Embarrassing. Not nice. Niceness can be a virtue. Most of us are nice people – we will not put other people out of countenance if we can help it, we will not deliberately rob them of comforting illusions. But niceness can be a dreadful vice as well. Here is what Heinrich Himmler had to say in a speech in 1943: 'In public we will never talk about it . . . It is with us, thank God, an inborn gift of tactfulness, that we have never conversed about this matter, never spoken about it . . . I am referring to the extermination of the Jewish people.'[12] Indeed Himmler did not refer, in public, to the 'extermination of the Jewish people': words like 'special treatment' were made to serve instead.

The Nazis, however, were not alone in their anxiety to avoid plain speaking. The victims did so too. We find certain elders of the Jewish community referring to the trains which took their brothers and sisters to the killing centres as 'favoured transport'; even in Auschwitz a crematorium would be called a 'bakery'.

And now? Who can avoid the parallel with our own situation, and the tricks of language which good taste and good form impose upon discussions – at least official discussions – of the Bomb? For 'special treatment' read 'demographic targeting', for 'bakery' read 'domestic fall-out shelter'.

But it would be wrong to pretend that it is only our own decent reticence which restrains us from breaking the taboos. We will – and we are generally quite well aware of it – be made to pay a public price for 'going too far'. When other people do not want to know, they *do not want to know* – and they do not take kindly to the self-appointed prophet who sees it as his duty to inform them.

In the old days it is said that kings would kill the messenger who brought bad news. Today, in the United States, the messenger may, as almost happened to the Plowshares 8, be silenced by the law; in Russia he may get locked up in a mental hospital. But there are other and subtler ways of restraining those who might otherwise speak out. And in our own country none is better tried or more effective than the technique of the social pillory. Anyone who forces an unwanted confrontation on the subject of the Bomb is liable to be punished for his impudence by being mocked, snubbed, made the butt of sneers and ridicule.

We all know the standard vocabulary of put-downs. 'Idealist', 'pacificist', 'moralist', 'holier-than-thou'. They have been with us a long time. The same schoolboy insults which Winston Churchill used to disparage those who objected to the idea of using anthrax bombs in the last war – 'psalm-singing defeatists' Churchill called them[13] – have now been dusted off by the brave editors of the *Daily This* or *Morning That* for use against our nuclear disarmers. And in settings where such clichés might be seen for what they are, cleverer men can be counted on to produce cleverer but equally dismissive sneers. Alistair Cooke, for example, writing about Bertrand Russell in the *Manchester Guardian* some years ago: 'A midget suspended against a huge Cinemascope screen . . . a charming puppet straining for a miracle and in the act wobbling the tiny wire frame of his body . . . "It is", he said in his high nasal voice, "the most important question men have ever had to decide in the whole history of the human race."'[14] A clever insult? Yes, a clever insult whereby a man who has shown too much emotion is defined as a puppet, a doll without emotions.

Lord Russell, peer of the realm and the greatest philosopher in England, was surely able to weather this kind of abuse. But few of us have Russell's social or intellectual confidence. Hardly surprising, then, if we sometimes persuade ourselves that, whatever we think privately, it is just not our place to make a public stand against the Bomb. Lords, philosophers, actresses, priests . . . they do that kind of thing. They can make exhibitions of themselves. But for the rest of us? Well, on the whole, all things considered, we like to keep calm. It is not our way to scream, or sing psalms, or call things the most important question in the whole history of the human race – even when the water is lapping round our feet. First one to panic is a wet.

But there is another and in some ways more telling reason why even the most courageous of us may be reluctant to speak out. Not so much that we mind the accusations of bad form, or that we are embarrassed to find ourselves mentioning an unmentionable problem, as that we are embarrassed to find ourselves mentioning it and yet *doing nothing more*. If we are going to alarm people we had better alarm them to some purpose; we had better offer a solution to the problem, and what is more we had better show by our example that we ourselves are actively pursuing it. We cannot simply knock on our neighbour's door and say, 'The world is standing on the brink of the final abyss . . . I thought you'd like to know.' If we ourselves do not have a solution, or if we are not prepared to dedicate our lives to finding one, then it is not only other people but our own consciences which will tell us to shut up. There is no honour whatever in being a *helpless* prophet – all dressed up with protest and nowhere to go.

Helplessness. I mean the dreadful feeling many of us know that there is in fact nothing we can do, that we are indeed midgets dwarfed by mighty forces over which individual human beings have no control.

I find no objective reasons for this helplessness. There is nothing in the political, economic or strategic situation which dictates that the world must continue on its present course. When people talk about the Russian threat, or about the power of the military–industrial complex, or about the unstoppable march of weapons technology, they are providing covers, not explanations, for why the race goes on. I have yet to hear of one good reason for not halting it tomorrow.

One good reason – except, that is, for the sense of helplessness itself. For helplessness can be a self-confirming process. It is a malady of the human spirit which once it has got going needs no good reasons to continue. When human beings *believe* themselves helpless, helpless they become.

Psychologists recognise two kinds of helplessness. *Learned* helplessness may develop when, for example, a person has repeatedly found that previous efforts to take control of his own life have genuinely come to nothing; he loses all faith in his own effectiveness and carries over to the present a picture of himself as someone unable any longer to influence events.

Does not learning play at least some part in the helplessness we now feel when confronted by nuclear weapons? We live in a society where people have in fact increasingly found themselves unable by their own efforts to take control over their lives. An unemployed labourer helpless to get himself a job, a homeless couple helpless to find themselves a lodging, a businessman helpless against market forces . . . Are such people likely to have faith in their power to act against the Bomb?

But there is also a different sort of helplessness: a *superstitious* helplessness whereby a person's belief in his own impotence has no basis in experience, but results instead from nothing more than a superstitious premonition that his life, and perhaps the life of the whole world, is set on an unalterable course – unalterable, that is, by human agency. The belief, for example, that his own fate has been sealed by a specific curse; or that, the world over, God and the Devil are working out their higher purposes without care for individual human beings. I say 'no more' than such a superstitious premonition – but superstitious helplessness can take the fight out of a man quite as effectively as any more reasonable fear. Cordelia Edvardson, one of the delegates to this year's reunion of Holocaust survivors, described how some of the Jews in Germany fell victim to just such a paralysing superstition: 'Of course', she says, 'we wanted to survive, but we were not at all sure we had the right to survive.'[15] And when a person no longer believes he has the right to survive, his helplessness itself is killing. I quote from a study of what has been called 'voodoo death': 'A Brazilian Indian condemned and sentenced by a medicine man dies within hours . . . In Australia a witch-doctor points a bone at a man. Believing that nothing can save him, the man rapidly sinks in spirits and prepares to die.'[16]

Earlier I cited the statistics from a *New Society* poll: nine out of ten people, worried by nuclear weapons, declared that there is nothing they can do. Nothing but sink in spirits and prepare to die? We behave at times as though we have been hexed by the Bomb, put under a spell.

A superstitious belief in the Bomb as an engine of fate over which human beings have no control has obvious origins in the human imagination. The Bomb is patently a superhuman weapon: mind-blowingly destructive, and, if we so see it, mind-blowingly magnificent. Small wonder if people's fear is mixed with awe – if they become hypnotised by the Bomb's dread beauty and its fascinating power.

The Bomb's first makers, the physicists who put it together in 1945, themselves treated their creation with almost mystical reverence. When Robert Oppenheimer witnessed the earliest test explosion in the New Mexico desert at Alamogordo, the words which came to him were from the holy book, the Bhagavad Gita:

> If the radiance of a thousand suns
> Were to burst at once into the sky
> That would be like the splendour of the Mighty One . . .
> I am become Death,
> The shatterer of worlds.

The test was given the code-name *Trinity*. And in the official report of the explosion the language was full of specifically Christian imagery. Here is part of the report which was rushed to President Truman, who was meeting with Churchill and Stalin at Potsdam:

> It lighted every peak, crevasse and ridge of the nearby mountain range with a clarity and beauty that cannot be described but must be seen to be imagined. It was the beauty the great poets dream about but describe most poorly and inadequately . . . Then came the strong, sustained, awesome roar which warned of doomsday and made us feel that we puny things were blasphemous to dare tamper with the forces heretofore reserved to the Almighty . . .[17]

Heretofore reserved to the Almighty – and hereinafter reserved to Truman, Churchill and Stalin? To Ronald Reagan, Margaret Thatcher and Yuri Andropov? No. We may be forgiven if we see our political leaders as servants, not masters to this force, and the Bomb itself as a wrathful giant which, having been woken from its slumber, will take vengeance on man's Jack-like arrogance.

It is easy to understand how people may move from this kind of superstitious image to a truly apocalyptic vision of a nuclear war. I mean now 'apocalyptic' in the old-fashioned sense: a Day of Judgement. A day when the Bomb will come to judge both the quick and the dead. A day which will be seen by some as a day of renewal, a cleansing holocaust – when our decadent civilisation must answer for its sins, its failure to understand or to make proper use of the gifts of science and technology, its failure in the Third World, its failure to establish a firm moral order.

For evidence that there are in fact people in this country who see the holocaust as a period of renewal, read the magazine of the nuclear shelter industry, *Protect and Survive Monthly*. The frontier spirit is, we'll find, still alive and well – and living 'somewhere in England' in ten years' time, when the survivors of the next World War will be leading heroically self-sufficient lives off the thin of the land, smiling and whistling and shooting their way out of all difficulties. The guns are 'primarily' for use against packs of marauding animals, though 'sadly' they may also be needed against our fellow human beings. Not that the next war won't be awful. It can and will be *wonderfully* awful – like Dunkirk, like the London Blitz – provided those whom they call 'defeatists, "moles", trendy clerics, cranks, fools and traitors' don't get their way and stop it happening.[18]

Or read – no, you will not be allowed to, because it's a confidential document – the Contingency War Plan of one of our Regional Health Authorities:

> In its way a nation is like a forest and the aim of war-planning is to secure survival of the great trees . . . If all the great trees and much of the brushwood are felled, a forest may not regenerate for centuries. If a sufficient number of the great trees is left, however, if felling is to some extent selective and controlled, recovery is swift . . . There will remain brushwood enough, if 30 million survivors may be so described. The planning policy is clearly elitist . . .[19]

Apocalyptic fantasies have always lurked beneath the surface of men's subconscious minds. They have emerged again and again in history at times of trouble, uncertainty, moral insecurity. In the Middle Ages the image of the Day of Judgement would have been familiar to us, one of the few pictures we knew, painted up above the transept on the walls of the local church. We would have heard the words of St John's Revelation thundered from the pulpit. I quote them now as a description of a fantasy which could be the reality of nuclear war:

> I looked, and behold a pale horse: and his name that sat on him was Death, and Hell followed with him . . . and, lo, there was a great earthquake; and the sun became black as sackcloth of hair, and the moon became as blood; And the stars of heaven fell unto the earth . . . And the heaven departed as a scroll when it is rolled together; and every mountain and island were moved out of their places. And the kings of the earth, and the great men, and the rich men, and the chief captains, and the mighty men, and every bondman, and every free man, hid themselves in the dens and in the rocks of the mountains; And said to the mountains and rocks, Fall on us, and hide us from the face of him that sitteth on the throne, and from the wrath of the Lamb . . .[20]

Fall on us – that strange imperative. In this state of mind people may come to welcome the very thing they fear, they may feel lured to the imminent catastrophe – like Trilby to her Svengali, like rabbits mesmerised by the twisting snake. Such is the Strangelove Syndrome. An attachment to the engines of destruction, an attachment even to the blissful state of being destroyed.

For people do not really accept the fact of their own death. Like a suicide who leaves a note, 'I picture you reading this note when I am gone', people picture themselves standing above the chaos in which they themselves have died – and may experience a sickening excitement at the images of destruction and decay.

'Do it beautifully,' says Hedda Gabler to Lovborg, as she hands him the gun. Oh yes, we'll do it beautifully. What more beautiful way to do it than in the way that poets dream about, but describe most poorly and inadequately? But the gun goes off by accident, and Lovborg dies miserably, shot not through the heart but through the balls. The Bomb is not beautiful. We must call down the curtain on this Tragic Play, scream it off the stage.

Silence, said Nadezhda Mandelstam, is the real crime. In Russian her name '*nadezhda*' means 'hope'. The hope lies in hope. Just as despair can be a self-fulfilling prophecy, so can its opposite. Hope, too, will create its own object – by giving us the strength of mind and voice to tackle our embarrassment, our helplessness, our own dark images of death, and come through to a world not merely of our making but of our choosing.

When Mountbatten asks 'Do any of those responsible for this disastrous course pull themselves together and reach for the brakes?' the answer must be 'Watch me!' And the answer 'No' can be reserved for a different question, the question Jacob Bronowski himself asked at the end of his essay, *Science and Human Values*: 'Has science fastened upon our society a monstrous gift of destruction which we can neither undo nor master, and which like a clockwork automaton is set to break our necks?' No. The Bomb is not an uncontrollable automaton, and we are not uncontrolling people.

Our control lies – as it always has done *whenever it's been tried* – in the force of public argument and public anger. It was public opinion in this country which forced the ending of the slave-trade – opinion marshalled then as it can be now by pamphlets, speeches and meetings in every village hall. It was fear of the public's outcry which prevented President Nixon from using an atom bomb in Vietnam, and it was the protests of the American people against that cruel and pointless war which eventually secured the American withdrawal.

We forget, sometimes, our own power. In this country every penny spent on armaments is money *we* subscribe, every acre of grass behind every barbed-wire fence round every bomber base is an acre of *our* land, and every decision taken by every Minister of State is a decision made on *our* behalf by a representative elected to *our* service. If those we entrust to manage our affairs adopt strange policies; if they turn out, in office, to be double agents – one hand to pat our babies, the other raised in salute to the Bomb – then we have the right and the duty to dismiss them as unfit.

What happens when an irresistible force meets a *movable* object? Why, it moves. But it will not happen quietly. Nadezhda's hope was loud and strident. Ours must be too.

Dylan Thomas spoke these words to his ageing father:

> Do not go gentle into that good night,
> Rage, rage, against the dying of the light.

REFERENCES

Chapter 1: Introduction: Homo psychologicus

1. A. S. Ramsey, *Dynamics*, Cambridge University Press, 1954.
2. Thomas Hobbes, *Leviathan*, ed. M. Oakeshott, Oxford University Press, 1946.
3. L. Wittgenstein, *Philosophical Investigations*, Blackwell, 1958, I, 294.
4. J. Dunn and C. Kendrick, *Siblings: Love, Envy, and Understanding*, Harvard University Press, 1982.
5. G. Lienhardt, *Divinity and Experience: The Religion of the Dinka*, Clarendon Press, 1961.
6. R. M. Pirsig, *Zen and the Art of Motorcycle Maintenance*, Bodley Head, 1974.

Chapter 2: The social function of intellect

1. G. M. French, 'Associative problems', in *Behavior of Non-human Primates*, ed. A. M. Schrier, H. F. Harlow and F. Stollnitz, Academic Press, 1965
2. G. G. Gallup, 'Chimpanzees: self-recognition', *Science*, **167**, 86, 1970.
3. N. K. Humphrey, 'Predispositions to learn', in *Constraints on Learning: Limitations and Predispositions*, ed. R. A. Hinde and J. Stevenson-Hinde, Academic Press, 1973.
4. A. W. Heim, *The Appraisal of Intelligence*, Methuen, 1970.
5. J. Goodall, 'Tool using and aimed throwing in a community of free-living chimpanzees', *Nature*, **201**, 1264, 1964; G. Teleki, 'Chimpanzee subsistence technology: materials and skills', *Journal of Human Evolution*, **3**, 575, 1974.
6. M. Sahlins, *Stone Age Economics*, Tavistock, 1974.
7. G. Teleki, op. cit. (note 5 above).
8. M. Sahlins, op. cit. (note 6 above).
9. N. K. Humphrey, 'Vision in a monkey without striate cortex: a case study', *Perception*, **3**, 241, 1974.
10. R. W. Wrangham, 'The behavioural ecology of chimpanzees in Gombe National Park, Tanzania', Ph.D. thesis, Cambridge, 1975.
11. C. Lévi-Strauss, *The Savage Mind*, Weidenfeld & Nicolson, 1962.
12. J. Bowlby, *Attachment and Loss*, Hogarth Press, 1969.
13. C. H. Waddington, *The Ethical Animal*, Allen & Unwin, 1960.
14. P. C. Wason and P. N. Johnson-Laird, *Psychology of Reasoning*, Batsford, 1972.
15. C. Trevarthen, 'Conversations with a two-month-old', *New Scientist*, **62**, 230, 1974.
16. E.g. J. Taylor, *Superminds*, Macmillan, 1975.
17. *New Scientist*, 67, 252, 1975.

Chapter 3: Nature's psychologists

1. K. J. W. Craik, *The Nature of Explanation*, Cambridge University Press, 1943.
2. N. Chomsky, Review of B. F. Skinner's *Verbal Behavior*, *Language*, 35, 26, 1959.
3. Cf. R. A. Hinde, *Animal Behaviour*, McGraw Hill, 1970.
4. L. Wittgenstein, op. cit. (Chapter 1, note 3 above), I, 293.
5. ibid., 272.
6. N. K. Humphrey, op. cit. (Chapter 2, note 9 above).
7. L. Weiskrantz, E. K. Warrington, M. D. Sanders and J. Marshall, 'Visual capacity in the hemianopic field following a restricted occipital ablation', *Brain*, 97, 709, 1974.

Chapter 5: Consciousness: a Just-So Story

1. L. Wittgenstein, op. cit. (Chapter 1, note 3 above), I, 293.
2. L. Weiskrantz and others, op. cit. (Chapter 3, note 7 above).
3. J. B. Watson, *Behaviorism*, Kegan Paul, 1928.
4. B. F. Skinner, 'The steep and thorny way to a science of behaviour', in *Problems of Scientific Revolution*, ed. R. Harré, Oxford University Press, 1975.
5. P. B. Porter, 'Another puzzle-picture', *American Journal of Psychology*, 67, 550, 1954.
6. D. Premack and G. Woodruff, 'Do chimpanzees have a theory of mind?', *Behavioural Brain Sciences*, **4**, 515, 1978.

Chapter 6: Joining the club

1. G. Ryle, *The Concept of Mind*, Hutchinson, 1949.
2. W. B. Yeats, quoted in C. Ricks, *Keats and Embarrassment*, Oxford University Press, 1976.
3. V. Turner, *The Forest of Symbols: Aspects of Ndembu Ritual*, Cornell University Press, 1967.
4. S. Rose, *New Scientist*, 21 September 1978.
5. E. Kübler-Ross, *On Death and Dying*, Tavistock, 1970.
6. A. Alvarez, *The Savage God: A Study of Suicide*, Weidenfeld & Nicolson, 1971.
7. B. Pasternak, *An Essay in Autobiography*, Collins and Harvill Press, 1959.
8. D. Magarshak, Introduction to Fyodor Dostoyevsky, *The Idiot*, Penguin, 1955.
9. Thomas Reid, *Essays on the Intellectual Powers of Man*, 1785; MIT Press, 1969.
10. P. Shuttle and P. Redgrove, *The Wise Wound: Menstruation and Everywoman*, Gollancz, 1978.

Chapter 7: Dreaming and being dreamed

1. J. S. Bruner, A. Jolly and K. Silva, *Play*, Penguin, 1976.
2. E.g. P. C. Reynolds, 'Play, language and human evolution', in J. S. Bruner and others, op. cit. (note 1 above).
3. Simone de Beauvoir, *Memoirs of a Dutiful Daughter*, quoted in 'Character games', ibid.

4. J. B. Loudon, 'Teasing and socialization on Tristan da Cunha', in *Socialization: the Approach from Social Anthropology*, ed. P. Mayer, Tavistock, 1970.
5. B. E. Ward, 'Temper tantrums in Kau Sai', ibid.
6. P. and I. Mayer, 'Socialization by peers: the youth organization of the Red Xhosa', ibid.
7. C. W. Kimmins, 'Children's dreams', in *A Handbook of Child Psychology*, ed. C. Murchison, Clark University Press, 1931.
8. ibid.
9. H. P. Roffwarg, J. Muzio and W. Dement, 'The ontogenetic development of the human sleep–dream cycle', *Science*, 152, 604, 1966.
10. St Teresa, quoted by G. Bataille, *Eroticism*, Calder, 1962.
11. S. Freud, *The Interpretation of Dreams*, Allen & Unwin, 1954.
12. ibid.
13. C. W. Kimmins, op. cit. (note 7 above).
14. C. Fisher, J. Gross and J. Zuch, 'Cycle of penile erection synchronous with dreaming (REM) sleep', *Archives of General Psychiatry*, 12, 29, 1965.
15. C. W. Kimmins, op. cit. (note 7 above).
16. Thomas Browne, 'On Dreams' (1650), in *Dreams and Dreaming*, ed. S. G. M. Lee and A. R. Mayes, Penguin, 1973.
17. C. W. Kimmins, op. cit. (note 7 above).
18. ibid.
19. C. G. Jung, *Man and His Symbols*, Aldus, 1964.
20. Mireille Bertrand, personal communication, 1979.
21. C. G. Jung, *The Archetypes and the Collective Unconscious*, Routledge & Kegan Paul, 1968.

Chapter 8: What's he to Hecuba?

1. C. W. Kimmins, op. cit. (Chapter 7, note 7 above).
2. Data from: A. H. Carding, 'The growth of pet populations in Western Europe', in *Pet Animals and Society*, ed. R. S. Anderson, Baillière Tindall, 1975; R. D. Godwin, 'Trends in the ownership of domestic pets in Great Britain', ibid.; R. Mugford, 'The contribution of pets to human development', unpublished review for Pedigree Petfoods, 1977.
3. K. Lorenz, cited by I. Eibl-Eibesfeldt, *Ethology*, Holt, Rinehart, Winston, 1970.
4. B. M. Levinson, 'Pets and environment', in R. S. Anderson, op. cit. (note 2 above).
5. G. Bateson, *Naven*, 2nd ed., Stanford University Press, 1958.
6. V. Turner, op. cit. (Chapter 6, note 3 above).
7. A. van Gennep, *The Rites of Passage*, English ed., Routledge & Kegan Paul, 1960.
8. K. E. Paige, 'The ritual of circumcision', *Human Nature*, May 1978.
9. R. Lambert, *The Hothouse Society*, Weidenfeld & Nicolson, 1968.
10. E. R. Leach, *Culture and Communication: The Logic by which Symbols are Connected*, Cambridge University Press, 1976.
11. ibid.
12. C. Lévi-Strauss, *The Raw and the Cooked*, Cape, 1970.

13. F. Nietzsche, *A Nietzsche Reader*, ed. R. J. Hollingdale, Penguin, 1977.
14. A. S. Byatt, 'Imagining the worst', *Listener*, 15 July 1976.
15. V. Turner, op. cit. (Chapter 6, note 3 above).
16. F. L. Lucas, *Tragedy: Serious Drama in Relation to Aristotle's Poetics*, Hogarth Press, 1957.
17. C. Stanislavski, *An Actor's Handbook*, Theatre Arts Books, 1963.
18. E. Goffman, *Frame Analysis*, Penguin, 1975.
19. Walter H. Pater, *Studies in the History of the Renaissance*, Macmillan, 1873.
20. E.g. R. A. Hinde, 'The concept of function', in *Function and Evolution of Behaviour*, ed. G. Baerends, C. Beer and A. Manning, Oxford University Press, 1975.

Chapter 9: The illusion of beauty

1. Thomas Reid, op. cit. (Chapter 6, note 9 above).
2. C. Bell, *Art*, Chatto & Windus, 1913.
3. W. Empson, *Seven Types of Ambiguity*, Chatto & Windus, 1930.
4. C. Lévi-Strauss, *Structural Anthropology*, Basic Books, 1963.
5. Johann Herbart, *Practical Philosophy*, 1808.
6. K. Pahlen, *Music of the World: A History*, Spring Books, 1963.
7. C. Humphrey, 'Some ideas of Saussure applied to Buryat magical drawings', in *Social Anthropology and Language*, ed. E. Ardener, Tavistock, 1971.
8. G. M. Hopkins, 'On the origin of beauty: A Platonic dialogue', in *G. M. Hopkins: Journals and Papers*, ed. H. House and G. Storey, Oxford University Press, 1959.
9. A. N. Whitehead, *An Enquiry Concerning the Principles of Natural Knowledge*, Cambridge University Press, 1919.
10. Aristotle, *Poetics* IV.
11. N. K. Humphrey and G. Keeble, 'How monkeys acquire a new way of seeing', *Perception*, 5, 51, 1976 (and earlier references ibid.).
12. D. C. McLelland, J. W. Atkinson, R. A. Clark and E. L. Lowell, *The Achievement Motive*, Appleton–Century, 1953.
13. J. Kagan, 'Attention and psychological change in the young child', *Science*, 170, 826, 1970.
14. P. P. G. Bateson, 'Internal influences on early learning in birds', in R. A. Hinde and J. Stevenson-Hinde, op. cit. (Chapter 2, note 3 above).
15. R. Wilbur, 'Poetry and the landscape', in *The New Landscape in Art and Science*, ed. G. Kepes, Paul Theobald, 1956.
16. A. P. Herbert, quoted in M. Hadfield, *The Gardener's Companion*, Dent, 1936.
17. I. P. Pavlov, 'The reflex of purpose', in *Lectures on Conditioned Reflexes*, vol. 1, Lawrence & Wishart, 1941.

Chapter 10: Turning the left cheek

1. V. Van Gogh, *The Complete Letters of Vincent Van Gogh*, 1958.
2. I. C. McManus and N. K. Humphrey, 'Turning the left cheek', *Nature*, 243, 271, 1973.
3. I. C. McManus provides a fuller analysis of this and related problems in 'Determinants of laterality in man', Ph.D. thesis, Cambridge, 1978.

Chapter 11: Butterflies that stamp

1. *Sunday Times*, 16 June 1974.
2. *Guardian*, 20 October 1973.

Chapter 12: The colour currency of nature

1. N. K. Humphrey, 'Interest and pleasure: two determinants of a monkey's visual preferences', *Perception*, **1**, 395, 1972.
2. N. K. Humphrey and G. Keeble, 'The reaction of monkeys to fearsome pictures', *Nature*, **251**, 500, 1974.
3. T. Porter, 'An investigation into colour preferences', *Designer*, September 1973.
4. R. M. Gerard, quoted by J. M. Fitch, 'The control of the luminous environment', *Scientific American*, **219**, 190, 1968.
5. K. Goldstein, 'Some experimental observations concerning the influence of colors on the function of the organism', *Occupational Therapy*, **21**, 147, 1942; L. Halpern, 'Additional contributions to the sensorimotor induction syndrome in unilateral disequilibrium with special reference to the effects of colors', *Journal of Nervous and Mental Diseases*, **123**, 34, 1956.
6. B. Wright and L. Rainwater, 'The meanings of color', *Journal of General Psychology*, **67**, 89, 1962.
7. E. Pinkerton and N. K. Humphrey, 'The apparent heaviness of colours', *Nature*, **250**, 164, 1974.
8. B. Berlin and P. Kay, *Basic Color Terms*, University of Los Angeles Press, 1969.
9. W. P. Brown, 'Studies of word listing', *Irish Journal of Psychology*, **3**, 117, 1972.
10. K. Goldstein, op. cit. (note 5 above).
11. N. K. Humphrey and G. Keeble, 'Effects of red light and loud noise on the rates at which monkeys sample the sensory environment', *Perception*, **7**, 343, 1978.
12. V. Turner, 'Colour classification in Ndembu ritual', in *Anthropological Approaches to the Study of Religion*, ed. M. Banton, Tavistock, 1966.

Chapter 13: Contrast illusions in perspective

1. M. Princet, quoted in C. H. Waddington, *Behind Appearance*, Edinburgh University Press, 1969.
2. From C. Blakemore and P. Sutton, 'Size adaptation: a new aftereffect', *Science*, **166**, 245, 1969.

Chapter 14: An ecology of ecstasy

1. O. Sacks, *Migraine: The Evolution of a Common Disorder*, Faber, 1971.

Chapter 16: Karma is raining on my head

1. K. R. Popper, 'Of clouds and clocks', in *Objective Knowledge*, Oxford University Press, 1972.
2. D. T. Campbell, 'Evolutionary epistemology', in *The Philosophy of Karl Popper*, ed. P. A. Schilpp, Open Court, 1974.

Chapter 19: Four minutes to midnight

1. J. Bronowski, *Science and Human Values*, Harper and Row, 1965.

2. *Guardian*, 15 January 1981.
3. Earl Mountbatten, 'The final abyss?', in *Apocalypse Now?*, Spokesman, 1980.
4. Lord Zuckerman, *Science Advisers, Scientific Advisers and Nuclear Weapons*, Menard Press, 1980.
5. *Guardian*, 25 May 1981.
6. *Financial Times*, 29 April 1981.
7. P. Payne, letter to *The Times*, 22 May 1981.
8. *New Society*, 25 September 1980.
9. Nadezhda Mandelstam, *Hope Against Hope*, Collins and Harvill Press, 1971.
10. A. Osada, in *Children of Hiroshima*, ed. Y. Fukushima, Taylor and Francis, 1980.
11. *New York Times*, 28 March 1981.
12. Quoted in Raul Hilberg, *The Destruction of the European Jews*, W. H. Allen, 1961.
13. *Listener*, 4 June 1981.
14. *Manchester Guardian*, 1955, quoted in E. P. Thompson, *Out of Apathy*, Stevens and Sons, 1960.
15. *Guardian*, 20 June 1981.
16. W. Cannon, quoted in M. Seligman, *Helplessness*, Freeman, 1975.
17. Brigadier Thomas Farrell, quoted in R. J. Lifton, *The Broken Connection*, Simon & Schuster, 1980.
18. *Protect and Survive Monthly*, 1981.
19. North East Thames Regional Health Authority Contingency War Plan, 1980.
20. The Revelation of St John the Divine, 6:8, 12–15.

INDEX

Page references in parenthesis indicate that an author's work is referred to on the page in question by a note (see references, pp. 211–16), without his or her name appearing in the main body of the text.